OXFORD STUDIES IN AFRICAN AFFAIRS

*General Editors*

JOHN D. HARGREAVES *and* MICHAEL TWADDLE

# CHARISMA AND BROTHERHOOD IN AFRICAN ISLAM

# Charisma and Brotherhood in African Islam

EDITED BY

Donal B. Cruise O'Brien

AND

Christian Coulon



CLARENDON PRESS · OXFORD
1988

*Oxford University Press, Walton Street, Oxford* OX2 6DP
*Oxford New York Toronto*
*Delhi Bombay Calcutta Madras Karachi*
*Petaling Jaya Singapore Hong Kong Tokyo*
*Nairobi Dar es Salaam Cape Town*
*Melbourne Auckland*
*and associated companies in*
*Berlin Ibadan*

*Oxford is a trade mark of Oxford University Press*

*Published in the United States*
*by Oxford University Press, New York*



*British Library Cataloguing in Publication Data*
*Charisma and brotherhood in African Islam.*
*1. Africa. Islam, 1780–1988*
*I. O'Brien, Donal B. Cruise II. Coulon, Christian*
*297'.096*
*ISBN 0–19–822723–X*

*Library of Congress Cataloging in Publication Data*
*Cruise O'Brien, Donal B. (Donal Brian)*
*Charisma and brotherhood in African Islam/Donal B. Cruise O'Brien and Christian Coulon.*
*(Oxford studies in African affairs)*
*Bibliography: p. Includes index.*
*1. Sufism—Africa, Sub-Saharan. 2. Islam and politics—Africa, Sub-Saharan. 3. Africa, Sub-Saharan—Politics and government—1960– 4. Leadership—Religious aspects—Sufism. I. Coulon, Christian. II. Title. III. Series.*
*BP64.A4S78 1989 297'.0967—dc19 88–12437*
*ISBN 0–19–822723–X*



*Typeset by Cotswold Typesetting Ltd, Gloucester*
*Printed in Great Britain*
*at the University Printing House, Oxford*
*by David Stanford*
*Printer to the University*

# Preface

THIS book is the first product of a four-year Franco-British collaborative research project, 'Islam, Society, and the State in Black Africa', jointly funded (1983–7) by the Economic and Social Research Council and the Centre National de la Recherche Scientifique. A research group of scholars, based in universities on the two sides of the Channel and under the co-direction of the editors of this volume, met twice yearly for the presentation and discussion of papers at each stage of presentation (draft outline, provisional text, final version). The aim of this project has been to produce two books: the second product is to be in French, *Les Communautés musulmanes et l'État en Afrique Noire*, to go to press in 1988.

The idea of this joint research project grew from a couple of international conferences in the School of Oriental and African Studies (1980, 1982) on the politics of African Islam. The research team which emerged from these meetings was drawn from various academic disciplines: political studies, history, anthropology, religious studies. The choice of the charismatic theme was perhaps a natural for such a divided scholarly community: but having made this choice of theme in our preliminary meeting, we have tried thereafter to follow it through consistently.

A certain habit of co-operation, as well as many firm friendships, have grown in the preparation of these studies. 'Franco-British', or Gascon–Irish, tensions have so far as possible been defused by the adoption of a Home Rule approach in the division of editorial tasks. The co-editorship of this book thus reflects a working reality, although it is the first editor who writes the Introduction. The same principle is hereafter to be applied for the second book, in French, where Dr Coulon's will be the principal responsibility.

It is a pleasure to express our thanks to the Centre National de la Recherche Scientifique and to the Economic and Social

Research Council for their generous funding. Individual authors below make separate acknowledgement for field research funding and for requisite academic leave. Thanks also to Mrs. Mette MacRae for her careful translations of the French contributions; to M. Jean Séguy, several times acknowledged below, for his stimulating contribution to our discussions; to the publisher's, readers, and editors for their very close and helpful critical reading of our texts. None of the above are to get any credit for our mistakes.

*D.C.O'B.*

# Contents

# List of Contributors

LOUIS BRENNER, Temporary Lecturer in African Religions, School of Oriental and African Studies, University of London

PETER CLARKE, Lecturer in the History and Philosophy of Religion, King's College, University of London

FRANCOIS CONSTANTIN, Professeur des Sciences Politiques, Université de Pau

CHRISTIAN COULON, Directeur Administratif, Centre d'Étude d'Afrique Noire, Université de Bordeaux I

DONAL B. CRUISE O'BRIEN, Reader in Politics with reference to Africa, School of Oriental and African Studies, University of London

MURRAY LAST, Lecturer in Anthropology, University College, University of London

JEAN-LOUIS TRIAUD, Maître Assistant, Departement d'Histoire, Université de Paris VII

RENÉ OTAYEK, Chercheur, Fondation Nationale des Sciences Politiques, Institut d'Étude Politique, Talence

# 1

# Introduction

## DONAL B. CRUISE O'BRIEN

IN the beginning was the miracle. The Muslim leaders dealt with in this volume, African saints and prophets of the past two hundred years, were all credited with miraculous achievements at decisive moments in their early exercise of leadership. The miracle is indeed to be seen as the decisive credential in this Islamic power-play. Thus a vision can be the basis for assuming supremacy in a new Sufi order, as with Sidi al-Mukhtar al-Kunti (1729–1811), or a miracle may intervene at the appropriate moment in a military campaign, as with al-Hajj ʿUmar Tal (1796–1864). The miracle may belong to the world of popular fantasy, as with Amadu Bamba's routing of his French gaolers (and hungry lions, a raging bull, snakes, and devils). The prodigious may also take a more modest and practical shape, as when Muhammad Jumat Imam, a Yoruba prophet, accurately predicts a rise in the price of a basic food stuff, *gari* (see Peter Clarke, in Chapter 8).[1]

The social climate of African Islam is one in which it is generally agreed that there can be extraordinary people, that the prodigious is also possible. Interpretation of the miraculous will indeed vary with the degree of Islamic learning in the audience: thus the erudite will see the miracle as the work of God alone, while the ignorant, blissfully unaware of the 'perils of associationism' (*shirk* or polytheism) will credit their hero, saint or prophet, with extraordinary, supernatural power. But the learned also accept that the extraordinary is possible, and that the miracle is a proof that the person in question has indeed been singled out by God.

A leader with such miraculous credentials has been in a position to recruit a large popular following. Disciples flocked to the man with a proven special access to divine favour, with

[1] The modesty of a miracle may in other circumstances be required if one is to avert the attentions of state authority, as among the Sufis of modern Egypt. See M. Gilsenan, *Recognizing Islam* (London: Croom Helm, 1982), 78–87.

supernatural powers. The miracle-working saint or prophet could bring supernatural assistance to bear in support of his disciples' worldly endeavours, whether in trade or in war. And the miracle was taken as proof that here was a leader who could intercede for his disciples on the Last Day, who could deliver on his promise of paradise in the hereafter. Success in business, success in war, with a guaranteed paradise as the bottom line: who might not have been tempted by such a charismatic contract?

The concern of these studies is firstly with the emergence of a particular style of exemplary leadership, saintly or prophetic, and then secondly with the development of Sufi Muslim institutions (brotherhoods) which trace their origin to such a miracle-working founder. The actors' idiom is that of the sacred, but the approach taken below often emphasizes the secular questions which are raised in the study of holy organization. Thus the saint or prophet is understood as a man of power, organizer of a mass of disciples around himself—sainthood is a form of personal rule in African Islam.[2] (The personal ruler has furthermore left a documented career, in the form of hagiography or holy polemic, an oral as well as a written record which allows a modern reconstruction in historical biography.)

The Islamic institution treated below is that of the Sufi brotherhood, instrument of power brokerage in a sacred code. In the study of African politics, any durable institution is perhaps in itself worthy of note. Around the sacred tombs of the founding saints we have a stable, valued, and recurring pattern of behaviour—under challenge now, but which has been present for nigh on two centuries.

The authors of the present studies decided, at an early stage in the preparation of their texts, to use Max Weber's concept of

[2] See R. H. Jackson and C. Rosberg, *Personal Rule in Black Africa* (Berkeley: University of California Press, 1982). The Muslim personal ruler with whom we are concerned has built and managed institutions, usually in the form of Sufi brotherhoods. The *Walī Allah* (friend of God) is the type of individual here presented as the saint. Such an equation of saint and *walī*, although current in the social sciences, can of course be questioned, as by J. Baldick, 'Among the Friends of God', *Times Literary Supplement*, 26 Sept. 1986.

charismatic authority as the agreed point of reference.[3] This was of course an obvious choice, although it did not command the unqualified support of every member of our research team. As a Franco-British collaborative research team, including representatives of four different university disciplines, and as many nationalities, perhaps we hoped that a German sociologist could provide us with a common language. At an early stage in our discussions we invited M. Jean Séguy, a French Weberian and authority on charisma in the medieval Catholic Church, to guide us in the use of this concept. M. Séguy was very courteous, read our prepared texts with attention, and apparently remained unconvinced that any of us was dealing with what should properly be termed charisma.[4]

Perhaps not, then: but it still seemed appropriate to use Weber's model as a measuring-rod for whatever it was that we were studying. The African Islamic ingredients, perhaps to be called 'not-quite-charisma' (NQC), seemed to conform to the Weberian ideal type in some basic respects: a miracle at the point of origin, recognized by a clientele in the grip of profound social crisis, saintly authority becoming traditional in the Sufi brotherhood. There were of course also to be problems in the use of a conceptual apparatus built far from the world of African Islam, but no doubt with various internal reservations it was agreed that we should persevere with charisma.

The Muslim concept of *baraka* (blessing, or gift of grace) was taken as (at least in certain instances) designating a power relation of the charismatic type. The miracle-worker, bearer of a personal gift (what Professor V. Crapanzano terms 'personal

[3] M. Weber, *Economy and Society*, ed. G. Roth and C. Wittich (New York: Bedminster, 1968). Portions of *Wirtschaft und Gesellschaft* (1925) had of course already been available in English translation, notably M. Weber, *The Theory of Social and Economic Organization*, ed. and trans. T. Parsons (New York: Oxford University Press, 1947), and in *From Max Weber: Essays in Sociology*, ed. and trans. H. Gerth and C. Wright Mills (New York: Oxford University Press, 1946). See also S. Eisenstadt (ed.), *On Charisma and Institution-Building by Max Weber* (Chicago: University of Chicago Press, 1968); B. S. Turner, *Weber and Islam* (London: Routledge and Kegan Paul, 1974).

[4] Jean Séguy, 'Charisme, sacerdoce, fondation: Autour de L. M. Grignion de Montfort', *Social Compass*, 29/1 (1982). Some of us noted that there was a '*Présence très limitée d'éléments miraculeux, au sens stricte, dans l'existence de L. M. Grignion de Montfort*', but we too were polite.

*baraka*') is the case in point.[5] This 'personal *baraka*' can be recognized in a variety of attributes: piety, spirituality, moral fibre, therapeutic gifts—Professor Clifford Geertz's 'moral vividness'.[6] *Baraka* is not of course identical with charisma: it is transmitted from one person to another, and it is inherited. Weber also in fact mentions the possibility of charisma being inherited or transmitted, along the path of routinization: in the case of hereditary charisma, the followers' recognition is no longer paid to the outstanding qualities of the individual leader, but to the legitimacy of the position which he has acquired by hereditary succession. The transmissible possibilities of charisma are also recognized by Weber, if in a limited number of cases—for the priest, in anointing or in the laying on of hands, and for the king, by anointing again, at the consecration. *Baraka*, which is communicated by touch (including the touch of a saintly tomb), may in this respect be seen as charisma in an Islamic idiom. A good traditional background (birth in the right sort of holy family) helps in the accumulation of *baraka*; personal *baraka*, of the kind which makes miracles possible is, however, properly seen as a gift from God.

When it is passed down as a birthright to all the descendants of a miracle-worker, *baraka* takes on an institutional character (Professor Crapanzano's 'institutional *baraka*'). In the Sufi brotherhood *baraka* still has the quality of grace, bearing the promise of salvation, but it also comes to be identified with worldly privilege, with wealth and power. The poor pay tribute in various forms to the bearers of grace, in the hope of a charitable return as well as of a passport to paradise. Thus the saint remains a source of comfort to the illiterate, and a pious visit to his tomb is the occasion of a communal investment in the afterlife.

Sufism or Islamic mysticism is required for the production of this Muslim charisma: one might indeed see Sufism as providing an Islamic handbook to the production of charisma. Birth and study help in the preparation of a charismatic personality, supplemented by pious retreat (*khalwa*), seclusion if possible in

[5] V. Crapanzano, *The Hamadsha: A Study in Moroccan Ethno-Psychiatry* (Berkeley: University of California Press, 1973).

[6] See C. Geertz, 'Centers, Kings and Charisma: Reflections on the Symbolics of Power', in *Local Knowledge* (New York: Basic Books, 1983).

the wilderness. And then may come the miracles, the dreams and visions, the revelations of sainthood. Muslim sainthood (*walāya*) actually seemed to provide a possible point of reference for our collaborative research, but it was discarded in favour of charisma after discussion. Some members of the team were without very plausible saints to consider, and Max Weber's conceptual schema was felt by most to have greater flexibility and range. (This is not true for all of our team: Louis Brenner manages very well in Chapter 2 without any mention of Weber or charisma. Sufism can thus provide an alternative approach to the study of leadership in Islam, another language of the charismatic.)

And then there were our doubts. Weber's charisma involves a devotion to 'the extraordinary and the unheard of, what is strange to all tradition'. Could this accurately be said of the Muslim leaders with whom we were concerned? The saint's or prophet's teaching may have taken novel forms, notably in concentrating popular devotion on the person of the leader in question, but it was also a teaching which usually emerged from a long tradition of local Islamic learning. The theologically ignorant could not become a saint, and the saints dealt with below were, at least in local terms, generally very learned. But then, to return to Weber, could not something similar be said in the case of Jesus Christ, surely the very archetype of the charismatic? Christ too was learned in the Jewish Torah: He had to know what was written before he could pronounce his innovating discourse.

Two other areas of doubt should also be mentioned, in relation to the absence of formal rules (abstract legal principles) in charismatic authority, and in relation to the anti-economic character of charisma—at least in Weber's 'pure type'. Where formal rules are concerned, even a miracle-worker in Islam remains aware of, and informed about, the legal texts of the *sharīʿa*. But there is a proverbial antinomy between the *sharīʿa* and the *ṭarīqa*, between the Islam of the legalist *ʿulamāʾ* (close to the courts of rulers) and the Islam of the *walī* and of the *ṭarīqa* (strong in rural retreat). Sufi arbitration of personal disputes, in the lodges of the brotherhoods, can have a flexible personal character absent from the dusty texts. Abstract legal principles are not absent from the world of Sufism, but neither are they obsessively invoked.

The Weberian idea that charisma spurns economic considerations, even if this is recognized to be more of an ideal than a fact, is difficult to fit with the Muslim cases dealt with below. Some saints were notably ascetic (Amadu Bamba, Shaikh Hamallah) but most were actively involved in trade (Sidi al-Mukhtar, Shaikh Sidiyya, Shaikh Ramiya), giving us an image of the saint as entrepreneur. Routinization of saintly authority in African Islam has led in some caes to an emphasis on commercial agriculture, as the disciples made a living from their calling. Muslim charisma (NQC) is heavily involved in worldly affairs, as it points to new economic as well as spiritual directions.

Weber's charisma did have the advantage of suggesting some pertinent questions. What were the inherent personal qualities which distinguished the leader? Why was he popularly perceived as extraordinary? And what were the social circumstances which helped to generate the charismatic response, the devotion born of distress and enthusiasm? In examining the cases presented below, one could be forgiven for feeling that these questions may never be fully answered. First, for the inherent qualities (the 'germ' of the charismatic), how can we know what were the personal attributes of a charismatic leader? Hagiography is after all our principle source, a record sustained by propagandist scribes and praise-singers, working from material which often owes its origin to the holy man himself. To put the matter bluntly, then, how are we to avoid the charismatic confidence trick? The local enemies of a given leader may in some cases (al-Hajj 'Umar Tal) have left a record which allows for historical balance,[7] but if the leader was fully successful (Sidi al-Mukhtar al-Kunti) there would be no surviving record other than what was essentially his own. Of the various personal qualities which are mentioned in the hagiographic record, however, one does appear with regularity: the Muslim charismatic figure has a taste, and a capacity, for study.

[7] The historian's art is well displayed in two recent texts on al-Hajj 'Umar Tal: D. Robinson, *The Holy War of Umar Tal: The Western Sudan in the Mid-Nineteenth century* (Oxford: Clarendon Press, 1985); S. N. Mahibou and J. L. Triaud (ed. and trans.), *Voilà ce qui est arrivé: Bayan ma Waqa'a d'al Hagg 'Umar al-Futi, Plaidoyer pour une guerre sainte en Afrique de l'Ouest au 19ᵉ siècle* (Paris: Éditions du CNRS, 1983).

A formidable studiousness, in the reading and writing of sacred texts, would seem to be necessary to the preparation of a charismatic career. Perhaps it is this quality of studiousness which provides us with what Weber called the 'germ' of charisma. Allah, in the proverbial Muslim phrase, never created an ignorant *walī*. A converse hypothesis, that charisma compensates for a lack of learning, has, however, been suggested for some other Islamic times and circumstances.[8]

Detection of the circumstances of social crisis, which Weber sees as a requisite backdrop to the emergence of the charismatic hero, can also pose some serious historiographic problems. The Muslim leaders in question have indeed presented the society around them as gripped by crisis, subject to unjust rulers, and permeated by unbelief, but then such a presentation is very much in the charismatic interest. How can the outside observer, or the social historian, be sure that this crisis corresponds to a demonstrable reality? There are indeed some suggestive indications of a profound crisis in many African countries in the period under consideration, although one should of course beware of retrospective reasoning: some sort of 'crisis' is a feature of any society, although only a few societies produce charismatic leaders.

The charismatic leader, at least on some occasions, seems to have worked hard to create his own crisis. An example here is the nineteenth-century reinterpretation of unbelief at the hands of the leaders of *jihād*. Beliefs and practices which had long been tolerated were now to be stigmatized as pagan—even in the case of a professed Islam. Africa had of course long been characterized by a more or less peaceful coexistence between Islam and a range of local or kin-based religious systems. The new voice of Muslim puritanism and intolerance begins to be heard towards the end of the eighteenth century. It is associated with charismatic leadership (or with NQC), and it is highly subversive of existing belief and practice: adoption of the puritan stance leads to war, within and between African Muslim societies, in

[8] For a review of charisma among the ignorant, see M. Magassouba, *L'Islam au Sénégal: Demain les mollahs?* (Paris, Éditions Karthala, 1985).

the series of *jihād* movements of the nineteenth century.[9] The charismatic leader creating his own crisis is thus a regular feature of the Muslim landscape in our period. The leader in question may present himself as no trouble-maker, merely one bent on the restoration of proper standards, in truth a conservative. Even a leader who asks to be recognized as both Mahdī and Messiah, like Muhammad Jumat Imam, would have liked to be recognized (by British colonial government) as a conservative—the sort of conservative who wreaks havoc with tradition, with what has always existed (in Ijebu-Ode, in the Yoruba case reviewed by Peter Clarke in Chapter 8).

Professor N. Levtzion, in a recent review of the eighteenth-century evidence,[10] has pointed to two sets of external influences at work upon West African Muslim societies, one from North Africa and mediated by the Sahara, the other from the Atlantic coast. The eighteenth century is of course crucial to our understanding of the generation (or germination) of Islamic charisma, as the period immediately preceding the public emergence of some of the first Muslim leaders of the charismatic type (Sidi al-Mukhtar al-Kunti and Shaikh ʿUthman b. Fudi, contrasted by Louis Brenner below). External influences in the North African case relate to the spread of Sufi mysticism, sometimes associated with Islamic revivalism: it is noteworthy here that the Muslim thirteenth century opened in the year AD 1785–6, as the beginning of a Muslim century is the time for

[9] See M. Last, 'Reform in West Africa: The *jihād* Movements of the Nineteenth Century', in J. F. Ajayi and M. Crowder (eds.), *History of West Africa* (London: Longman, 1985); also M. Klein, 'Social and Economic Factors in the Muslim Revolutions in Senegambia', *Journal of African History*, 13/3 (1972); L. G. Colvin, 'Islam and the State of Kajoor: A Case of Successful Resistance to *jihād*', *Journal of African History*, 15/4 (1974). A special debt is to be acknowledged here to the many works of Professor J. S. Trimingham on the subject of African Islam: *Islam in the Sudan* (Oxford: Clarendon Press, 1949); *Islam in Ethiopia* (London: Oxford University Press, 1952); *Islam in West Africa* (Oxford: Clarendon Press, 1959); *A History of Islam in West Africa* (London: Oxford University Press, 1962); *Islam in East Africa* (Oxford: Clarendon Press, 1962); *The Influence of Islam upon Africa* (London and New York: Longman and Librairie du Liban, 1980).

[10] N. Levtzion, 'The Eighteenth-Century Background to the Islamic Revolutions in West Africa', conference on *Eighteenth-Century Renewal and Reform Movements in Islam* (Hebrew University, Jerusalem, 23–7 June 1985), published in N. Levtzion and J. O. Voll (eds.), *Eighteenth-Century Renewal and Reform in Islam* (Syracuse: Syracuse University Press, 1987).

expectation of a renewer (*mujaddid*) who will announce the coming of the Mahdī. The mood of Mahdist excitement may in particular have helped in creating a charismatic constituency for Sidi al-Mukhtar al-Kunti (1729–1811).

The second set of 'external influences' traced by Professor Levtzion are more widely known, namely, those relating to the still highly controversial issue of the effects of the slave-trade—transatlantic, Indian Ocean, or trans-Saharan—on African societies. The eighteenth century sees the transatlantic trade at its height. Destabilization in Muslim societies in West Africa, attendant upon the transatlantic trade, is strongly suggested by the pattern of the *jihād* movements—first developing close to the Atlantic coast, in areas already much affected by trade with Europeans (Futa Toro, Futa Jallon), and only erupting in the Hausa states in the nineteenth century after that area also began to be seriously affected by the Atlantic slave-trade (from *c.*1780, by Dr Fisher's recent estimate).[11] The pattern of an Atlantic influence (or one related to the slave-trade) does not, however, always fit: thus Professor Charles Stewart has highlighted the development of *jihād* in areas supplying grain to the Sahara.[12] The first case of Muslim charismatic leadership dealt with below, that of Sidi al-Mukhtar on the Saharan reaches of the Niger bend, occurs at a time when there is no (documented) involvement of that area with the distant Atlantic slave-trade.

The social disruption of that trade can however be argued with many other cases, as with the background to the *jihād* (and

[11] H. Fisher, 'A Muslim William Wilberforce? The Fulani *jihād* as Anti-Slavery Crusade', Islam seminar, School of Oriental and African Studies, 5 March 1985. For the suggestion of *jihād* movements following the impact of the Atlantic slave trade, see N. R. Keddie, 'Revolutionary *jihād* Movements in the Eighteenth and Nineteenth Centuries: Comparative Reflections', conference on *Eighteenth-Century Renewal and Reform Movements in Islam* (Hebrew University, Jerusalem), published in Levtzion and Voll (eds.), op. cit. (above, n. 10). See also J. R. Willis (ed.), *Slaves and Slavery in Muslim Africa*, 2 vols. (London: Frank Cass, 1985). M. Last, in the forthcoming revised edition of *History of West Africa* ('Reform in West Africa: The *jihād* Movements of the Nineteenth Century', pp. 8–10), suggests that the *jihād* movements are to be seen against a background of general crisis in the Fulbe pastoral economy, a crisis in which the Atlantic slave trade is of no more than partial import.

[12] C. Stewart, 'Southern Saharan Scholarship and the *Bilad al Sudan*', in *Journal of African History*, 17/4 (1976), 73–93.

the career) of 'Uthman b. Fudi.[13] The Shehu is thus presented by Dr Humphrey Fisher as preaching in the shadow of the Atlantic trade, his objection having been to the enslavement and sale of 'free Muslims'. Such slaves would furthermore, as his son Muhammad Bello stressed, fall into infidel, Christian hands. The institution of slavery was not in question for these jihadists; Muslims had after all done well enough out of the eighteenth-century trade, taking advantage of the great supply of slaves. Farming with slave labour was a crucial element in Fulbe sedentarization, a feature of the post-*jihād* states of the Sokoto Caliphate. In overall terms, one may perhaps say that Muslims were more enslavers then enslaved: the effects of the Atlantic trade, while destructive of some societies, helped in the construction of others.

If the effects of the long-distance slave-trade were probably an important part of the background to Muslim upheavals in Black Africa, then the abolition and suppression of that trade in the nineteenth century was also to provide a background of social crisis appropriate for the emergence of charismatic leadership in Muslim societies. Unemployment among the slave-gatherers and the gradual liberation of slaves were to help in the provision of a charismatic clientele, an audience ready to hear the message of a new Muslim leader, ready to believe in a miracle. There is here a significant connection between Muslim charisma and the development of commercial agriculture for sale to the Europeans (Shaikh Sidiyya trading gum arabic to the Senegal river, Amadu Bamba's Mourides selling ground-nuts to Dakar, Shaikh Ramiya's disciples selling copra to Bagamoyo).[14]

Reasoning in terms of a social crisis, at least in overall terms, would seem to suggest socio-economic circumstances of the sort

[13] D. M. Last, *The Sokoto Caliphate* (London: Longman, 1967); M. Hiskett, *The Sword of Truth: The Life and Times of the Shehu Usuman dan Fodio* (New York: Oxford University Press, 1973); L. Brenner, 'Muslim Thought in Eighteenth-Century West Africa: The Case of Shaikh Uthman b. Fudi', conference on *Eighteenth-Century Renewal and Reform Movements in Islam* (Hebrew University, Jerusalem), published in Levtzion and Voll (eds.), op. cit. (above, n. 10). It may be noted that Shaikh 'Uthman denied that he was a saint, and that miracles were played down during his lifetime. But his descendants, notably his daughters, ensured that proper recognition was given to his achievements.

[14] D. B. Cruise O'Brien, *The Mourides of Senegal* (Oxford: Clarendon Press, 1971); A. H. Nimtz, *Islam and Politics in East Africa* (Minneapolis: University of Minnesota Press, 1980); C. Stewart with E. K. Stewart, *Islam and Social Order in Mauritania* (Oxford: Clarendon Press, 1973).

one might expect in the background to charismatic movements: the climax of the Atlantic slave-trade (eighteenth century), then its abolition and the development of 'legitimate commerce' (towards the mid nineteenth century), with all the attendant disturbances of social order. But there are still some obvious problems with the 'social crisis' approach. A crisis in these large terms is fairly general: why then should a charismatic response have been forthcoming only in some cases, and not in others? Can one confidently assert that crisis was less severe in cases when a charismatic leader did not emerge? We should at least attempt to isolate the sort of social crisis which has been generative of charismatic leadership. The collapse of state structures, as for example among the Wolof in the nineteenth century, is thus often part of the background to the development of Muslim charismatic movements. It can perhaps further be argued that such state structures or social units may have collapsed the more readily when they had already been weakened by being previously subsumed under an imperial power (Mali or Songhay in West Africa, Ottoman in North Africa, Zanzibari in the east).[15] The Wolof, in any case, certainly do provide an important special case for the development of brotherhood structures under charismatic leadership: the French in this instance, being largely untroubled by the religious allegiances of their imperial subjects, later even connived in the consolidation of Sufi brotherhoods as informal agencies of colonial rule.[16]

If the chosen time of Muslim sainthood is one of social and political crisis, then the chosen place might seem likely to be one where political authority is almost absent and where social structure is most dislocated. But where the charismatic place is concerned, it is not easy to reach any overall conclusion. Some saints, such as Sidi al-Mukhtar al-Kunti, did indeed develop their new sacred communities in the virtual absence of

[15] The imperial contribution to the breakdown of social structures, and thus to the subsequent emergence of charismatic leadership, was suggested in conversation by Nehemiah Levtzion (London, July 1986). See N. Levtzion, *Ancient Ghana and Mali* (London: Methuen, 1973); also M. Klein, *Islam and Imperialism in Senegal* (Edinburgh: Edinburgh University Press, 1968).

[16] C. Harrison, *France and Islam in West Africa 1860–1968* (Cambridge University Press, forthcoming); D. B. Cruise O'Brien, 'Chiefs, Saints and Bureaucrats', in *Saints and Politicians* (Cambridge: Cambridge University Press, 1975).

established political power, but others, such as Shaikh 'Uthman, worked in the shadow of political power (of the Hausa states in this instance).

The example of the Mossi people (in colonial Upper Volta, today's Burkina Faso, as studied by René Otayek in Chapter 5) provides some suggestive material on the geographical and historical distribution of Muslim charisma. The corporate cohesion of pre-colonial Mossi society, with its respected hierarchies and its traditional religious beliefs, left no room for Islamic charisma: and one might add that in intellectual terms the absence of Sufism meant the absence of a context for the production of saints. The decentralized societies around the fringes of the Mossi area, whether 'acephalous' or segmentary in organization, were hostile to the very idea of individuals with extraordinary power: such societies were in their way well organized, and they left no room for the emergence of individuals with the power of the extraordinary.

Islamic charisma was present among one group neighbouring the Mossi, the Fulani of Jelgooji. These pastoralists had long been Muslim: they conformed to the charismatic norm in that they inhabited a frontier zone, between the Mossi and the Muslim state of Macina. (In a seventeenth-century Moroccan case of Muslim sainthood, presented by Dr Hassan Elboudrari, we similarly find charisma thriving in the zone between the strong tribal groups of the mountains and the strong Spanish forts of the coast.)[17] Fulani allegiance to Hamallism[18] is interpreted below by René Otayek as the expression of a will to spritual independence (surrogate to the political, a sacred proto-nationalism), the disciples flocking to set up villages under the direction of their Hamallist guides. The establishment of these new villages is another recurrent theme in Muslim

[17] H. Elboudrari, 'Quand les saints font les villes: Lecture anthropologique de la pratique sociale d'un saint marocain du XVII^e siècle', *Annales ESC* (May–June 1985).

[18] The Hamallist brotherhood, an offshoot of the Tijāniyya, was founded by Shaikh Hamallah (*c.* 1883–1943) and became identified with anti-French opinion in colonial Sudan (today's Mali). See A. Traoré, *Islam et colonisation en Afrique: Cheikh Hamahoullah, homme de foi et résistant* (Paris: Maisonneuve et Larose, 1983); also L. Brenner, *West African Sufi: The Religious Heritage and Spiritual Search of Cerno Bokar Saalif Taal* (London: C. Hurst, 1984); P. Alexandre, 'A West African Islamic Movement: Hamallism in French Africa', in R. Rotberg and A. Mazrui (eds.), *Protest and Power in Black Africa* (New York: Oxford University Press, 1970).

charismatic movements, here as in many other cases conducted in conscious imitation of the flight of the Prophet Muhammad from Mecca to Madina, the *hijra*. Finally, one notes that these Hamallist Fulani were near enough to some major Muslim centres to maintain the holy learning that seems to have been indispensable to the nurture of Islamic charisma.

In the geographical distribution of charisma, the Muslims of the Mossi fringe may be compared with the Yoruba Mahdists studied by Peter Clarke in Chapter 8, who occupy yet another kind of interstitial position, forming a Muslim outpost in a society between rival belief systems, traditional religion and Christianity. The Mahdī in question appears to have declared himself such (as well as Christian Messiah) because he had not been made a Chief *Imam*—surely an extreme expression of a promotion grievance. A case such as this brings to mind a haunting problem in the study of charismatic movements: what about the failures? May we presume that for every success story, more or less well documented, there may be ten cases of abortive charismatic missions, which will be difficult if not impossible to discuss or document? Our understanding of Muslim charismatic movements would be enormously enhanced by a review of such failures: perhaps, for example, an inclination to travel, as reported by oral tradition, may conceal a record of rebuff and disappointment. The study of charismatic failure, however, goes beyond the pretensions of the present volume; it may be modestly—or perfidiously—presented as a challenge for future research ('Please tell me about your failures . . .').

Comparative research in the area of charismatic success stories, as essayed below, has led to a recognition of a knowing and willed quality in the pursuit of charismatic power. The aspirant saint appears to have been well aware of the rewards of holiness; he might even have contemplated sainthood as a career. But perhaps we have no right to be surprised. When a gifted student takes to his holy books, he is likely to be aware of the possible future rewards of study: in the world of Sufism, as elsewhere, knowledge may provide a route to power—a power extending in this case from the mystical to the political. This individual pursuit of power is then to be related to the surrounding social circumstances; here one must recognize the contribution of the charismatic clientele, the democratic or populist underpinnings of extraordinary power. All those involved would vehemently proclaim the irrelevance of these latter considerations, and insist that it is the leader (and the divinity, of course) who accomplishes everything. The incredulous outsider may, however, continue to suspect that the miracle owes much to the popular imagination, that the charismatic leader has been in an unusual sense popularly elected.

The relevant social circumstances in the generation of African Islamic charisma have also included the actions of European outsiders: attention is drawn, in the cases reviewed below, to the effects of European commercial penetration and European colonial policies in the unwitting stimulation of such charisma. Islam did greatly profit from the indirect and unintended results of colonial modernization in sub-Saharan Africa, making (in Professor Hiskett's estimation) many more conversions in some seventy years of colonial rule than it had done in the nine preceding centuries.[19] The colonial peace and improved communications (road and rail) in tropical Africa were greatly to facilitate the work of Muslim proselytism, carried onward by traders and by Sufis. The early European governments of African territories, in the period from 1890 to 1914, would not have desired any such bizarre result of their civilizing mission. The French and British governments were most apprehensive of the 'peril of Islam', the danger of a pan-Islamic rising sponsored by the Caliphate in Istanbul. It was for this reason that some of Africa's most charismatic Muslim leaders were particular objects of European mistrust and repressive action, as with the French imposition of long sentences of exile upon Amadu Bamba of Senegal (exiled 1895–1907) and Shaikh Hamallah of Sudan (exiled 1910–35). French power in these two cases, however, helped in the construction of Islamic charisma, and colonial repression had a major unintended outcome in leading many Africans to become members of the Mouride or Hamallist brotherhoods; here we have examples of the Sufi *ṭariqa* as anti-colonial party. These two cases may be taken to represent an extreme, a situation where African Muslim charisma comes close to what Weber calls the 'pure type' (no longer NQC?).

Where routinized charisma was concerned, as with the Sufi

[19] M. Hiskett, *The Development of Islam in West Africa* (London: Longman, 1984), 281.

establishment flourishing later both under British indirect rule in Northern Nigeria and under French (and only nominally direct) rule in Senegal, the colonial government could be altogether more accommodating. The saint's successors could then work effectively with the erstwhile colonialist enemy, securing a sacred area around the saintly tomb (as at Sokoto in Nigeria or Touba and Tivouane in Senegal). The colonial period could thus be a good time for the Sufi brotherhood, as it could also be for the Mahdī who came to announce the world's end: the Mahdīs could benefit from the relative tolerance of a gaol sentence under colonial rule, whereas they might have been expected to meet with death at the hands of any Muslim government.

Returning to the notion of a charismatic career, of a craft for the creation of charisma (or for the cultivation of extraordinary power), one begins with the pursuit of holy learning. Literacy alone, or even a plausible semblance of literacy, had been enough to win a reputation for mysterious power while the surrounding population remained illiterate. The erudition of the Sufi saints who are dealt with in this volume belongs to a subsequent transitional period in African Islam, when Arabic literacy was becoming more widespread. The relevant knowledge consists first, of course, in a thorough mastery of the contents of the Qur'ān, then of the authorized traditions, and perhaps of the specialized disciplines of instruction. Knowledge is then also a self-knowledge, entering the domain of mystical experience by the paths of numerology and divination;[20] it is probably significant that many charismatic Muslim leaders are self-educated. Some, such as Shaikh Ramiya of Tanganyika (1856–1931), might not have been regarded as impressively learned if set in another context; but, as François Constantin rightly remarks in Chapter 4, it is Ramiya's learning in the context of the East African hinterland that helps account for his reputation and power.[21]

[20] L. Brenner, 'The Esoteric Sciences in West African Islam', in I. H. Abdalla and B. M. Du Toit (eds.), *African Healing Strategies* (Buffalo, NY: Trado-Medic Books, 1985); also id., *Réflexions sur le savoir islamique en Afrique de l'Ouest* (Bordeaux: Centre d'Étude d'Afrique Noire, 1985).

[21] See also J. L. Triaud and F. Constantin (eds.), *Les Leaders religieux islamiques en Afrique Orientale* (Paris: Credu/Karthala, forthcoming).

Among some of the newly Muslim peoples of sub-Saharan Africa, Arabic literacy was by the nineteenth century no longer enough for the individual in pursuit of the extraordinary. As Arabic literacy spread, so did the special disciplines of Islamic mysticism come to the fore in the preparation of charisma: as Murray Last remarks in Chapter 9, the recipe for charisma had been changed. Now it was Sufism which offered the special secrets necessary for personal power. The religious retreat or *khalwa*, as described by Jean-Louis Triaud in Chapter 3, is particularly important in this context. The practice of *khalwa* (for periods of ten, twenty, thirty, or even forty days) is an exercise in mastery, not only of hunger and sleep, but also of the terrors of solitude and the supernatural. Such a cultivation of the extraordinary is seen as part of a well-defined course (like a difficult university examination) to be taken by the aspirant to the Sufi title of *shaikh.* The practice of religious exercises of this kind, demonstrations of exemplary piety, could also be undertaken as preparation for the moment of spiritual opening (*fatḥ*) by which came the divine gift of sainthood. And with *walāya* or sainthood was then to come a spiritual clientele, mastery indeed, as miracles and visions provided the decisive proof. (However, the voice of scepticism is also heard below, as François Constantin sees in such miracles some remarkably successful manipulations.)

The notion of charisma as a career, a matter of personal choice for the ambitious, is exemplified in Louis Brenner's account in Chapter 2 of Sidi al-Mukhtar al-Kunti. Sidi al-Mukhtar was very well educated in his celebrated Saharan family, the Kunta of Timbuktu, and became convinced of his own sainthood by the age of nineteen. His first job, as guardian of a saintly tomb (Ahmad al-Bekkai), allowed him to establish the personal relations which he went on to use in building his own religious brotherhood (the Qādiriyya-Mukhtāriyya). He was determined to build this community around himself, claiming that it represented 'the only pure Islam in his time'. And sainthood did not inhibit him from using 'every means at his disposal to overcome his competitors and adversaries, including diplomacy, sharp business practice, [and] employing the awesome power of his curse'. The title 'Head of the Brotherhood' (*shaikh al-ṭarīqa al-Qādiriyya*) he appears to have awarded himself; he used the brotherhood networks to extend and cement his kinship relations, while his sainthood served to reinforce his authority as neutral arbiter in disputes. One might add that he also apparently used his own wealth to settle the claims in intractable disputes, thus building his reputation for holiness. Whatever the secrets of success, there is no doubt that his was a particularly impressive Saharan career.[22]

The use of charisma as a political instrument is again apparent in the case of Nigeria at the time of the early competition between the rival Sufi brotherhoods Tijāniyya and Qādiriyya. The Qādiriyya as a mystical association, linked into the life of Shaikh ʿUthman b. Fudi, had a shadowy pre-eminence in Hausaland from the late eighteenth century on. But it was only through the subsequent stress of competition with a rival brotherhood, in contesting the exclusivist Tijāni claims to spiritual superiority, that the Qādiriyya took clear political shape as an organized brotherhood. This development can be dated from 1837, with the death of Shaikh ʿUthman's son Muhammad Bello. A bitter controversy developed after Bello's death as to whether or not he had accepted an initiation to the Tijāniyya. If he had done so, the necessary implication was that he had rejected the Qādiriyya, the brotherhood of the political élite. The Qādirī establishment then counter-attacked, promoting the Qādiriyya as the unique brotherhood and Shaikh ʿUthman as the best of saints. His miracles were publicized, together with promotion of the idea that his tomb should be a second place of pilgrimage after Mecca. This controversy between brotherhoods, which has lived on in northern Nigeria

[22] On building a saintly reputation by personal contribution to mediation settlements see E. A. Macdougall, 'The Economics of Islam in the Southern Sahara: The Rise of the Kunta Clan', in H. J. Fisher and N. Levtzion (eds.), *Rural and Urban Islam in West Africa*, published as *African and Asian Studies*, 20/1 (Mar. 1986), 56. See also A. A. Batran, 'Sidi al-Mukhtar al-Kunti, and the Recrudescence of Islam in the Western Sahara and the Middle Niger', Ph.D. thesis (University of Birmingham, 1971), and id., 'The Kunta, Sidi al-Mukhtar al-Kunti, and the Office of *Shaykh al-Ṭarīqa al-Qādiriyya*', in J. R. Willis (ed.), *Studies in West African Islamic History*, i. *The Cultivators of Islam* (London: Frank Cass, 1979). On the Kunta background see also J. R. Willis, 'The Western Sudan from the Moroccan Invasion (1591) to the Death of Sidi al-Mukhtar al-Kunti (1811)', in J. F. A. Ajayi and M. Crowder (eds.), *History of West Africa*, i (Longman, 1985), and E. Saad, *Social History of Timbuktu* (Cambridge: Cambridge University Press, 1983). For an appraisal of the Kunta role in Islamic leadership see Stewart, *Islam and Social Order in Mauritania*, pp. 36–45. Quotations in this paragraph are from L. Brenner in Ch. 2 below.

to the present day,[23] exemplifies the politicization of brotherhood links.

A modern instance of the charismatic career is provided in Chapter 6 in the case of a Senegalese woman *shaikh*, Sokhna Magat Diop, who has inherited her *baraka* from both mother and father. This Mouride *shaikh* never attends the mosque, man's domain, but concentrates instead on the practice of pious retreat (*khalwa*), through which she maintains her authority over numerous disciples—male as well as female. And Christian Coulon, in providing this account, asks whether mysticism, in blurring sexual differences, could be the instrument of feminine authority in Islam.

Of course, charisma also has its democratic or populist dimension, in the recognition which the disciples extend to their leader. It is not enough to point to saintly manipulations, or indeed to call on God. Why do the charismatic devotees choose to accept the miraculous? Perhaps part of the answer is that charisma draws its strength from people's anxieties. Thus the spread of popular fear about the afterlife in the nineteenth century, with the diffusion of the notion of punishment after death, appears to have accompanied the growth in the popular audience of the *walī*. The saint in such an instance is seen as having helped to create the anxiety which he could best assuage. The charismatic Muslim leader here belongs to a period when people have passed over a certain threshold of Islamization, and are in the process of becoming fully Muslim.

A populist character has been recognized in the Sufi Muslim tradition, in the provision for a 'warm' and personal piety, focused on devotion to an exemplary human being, the saint. Such an Islam, in North Africa for example,[24] has long had an appeal to the lower orders in society, to the illiterate, the artisans, and the shanty-towners, in contradistinction to the scripturalist Islam of high society and the court. Saint-worship

[23] See J. Paden, *Religion and Political Culture in Kano* (Berkeley: University of California Press, 1973); also id., *Ahmadu Bello: Sardauna of Sokoto* (London: Hodder and Stoughton, 1986).

[24] On North African Sufism see among others E. Gellner, *Saints of the Atlas* (London: Weidenfeld and Nicolson, 1969); E. Evans-Pritchard, *The Sanusi of Cyrenaica* (Oxford: Clarendon Press, 1949); C. Geertz, *Islam Observed: Religious Development in Morocco and Indonesia* (New Haven: Yale University Press, 1968).

can thus be a form of veiled popular opposition to current authority, which answers by denouncing it as heresy.

Popular demand for charismatic leadership is well illustrated by the example of East Africa, where black Africans of the hinterland had for centuries been effectively denied access to the world of Islam by the Arab and Swahili traders of the coast ('those who look towards the sea'). The Sufi brotherhood (Qādiriyya again, or Shādhiliyya) in this case allowed black Africans of the interior to build an Islam of their own, with a saintly leader of their own country and their own race.[25] The British abolition of the slave trade in the early nineteenth century, by depriving the Arab sultans of their economic base and so destabilizing their political authority, offered the hinterland Africans their chance, which they seized by rallying to the cause of such men as Shaikh Ramiya. In this Tanganyikan case the devotees may be said to have elected their saint, in their eagerness for admission to the world of Islam and for recognition as the spiritual equals of their erstwhile masters from the coast. Such can be the democracy of charisma and the miracle.

Where popular Islam is concerned, one must of course bear in mind Africa's plurality of religious observances. Competition with traditional 'pagan' religion is a recurring theme in many of these charismatic cases, from Sidi al-Mukhtar al-Kunti's struggles against the opposing lineages of the desert at the end of the eighteenth century to the competition between Muslim healers and those of the Hausa *bori* cult in today's Nigeria. Concentration on Islamic themes should not lead us to disregard the conditions of religious pluralism which often exist around charismatic communities, and which lend particular colour to each local situation. The individual may accept elements of more than one religion, may believe apparently contradictory things. The domain of magic is important here and is explored in Chapter 9 by Murray Last in the field of healing in Nigeria (including the sinister power inherent in waistcoat buttons made of babies' eyes). The Muslim saint's appeal to the people, to the illiterate and the semi-literate as well as to the learned, includes a promise that the future can be known and

[25] Nimtz, *Islam and Politics in East Africa*.

controlled—a magical resource. And the saint may have a particular appeal to the weak and the socially marginal, to women, as notably in the Niass Tijāniyya, and to ex-slaves.

The European outsider continues to play his very significant role in stimulating the development of this charismatic Muslim leadership, first as the agent of rapid economic change (resulting from the transatlantic slave trade and its abolition) and then (with the imposition of colonial rule at the end of the nineteenth century) as the agent of an insecure and often unwelcome government. Colonial rule also brought the imposition of new commercial demands (the demand for African agricultural products being the most relevant here) which were to generate further social changes—and a further rapid progress of Islam. Improved communications allowed Muslim traders and preachers to penetrate into previously pagan areas, and the establishment of the colonial peace prevented the pagans from an armed exclusion of these Muslims. For such reasons one may say that where much of Africa is concerned the 'expansion of Europe' was also, and paradoxically, the expansion of Islam.

Some Muslim charismatic leaders (e.g. Amadu Bamba, Shaikh Hamallah) were also able to draw strength from the persecution which they endured at the hands of the colonial government. An element of 'sacred nationalism', of displaced anti-colonial resistance, may indeed be discerned in such cases, although the historian may remark that each of these two Muslim heroes fell victim to African plots as well as to European power (Hamallah to the intrigues of a rival brotherhood, the Tijāniyya, and Amadu Bamba to the intrigues of the Wolof chiefly party).[26] Furthermore, colonialist persecution did not always help, and anti-colonialism is not to be seen as automatically linked either to Sufism or to sainthood; many Sufis were to work closely with the Europeans, profiting mightily in their intermediary position between alien power and the uneasily governed masses.[27]

[26] The chiefly plot against Amadu Bamba in Cruise O'Brien, *The Mourides of Senegal*, pp. 40–4; the Tijāni plot against Shaikh Hamallah in Brenner, *West African Sufi*, pp. 131–8.

[27] On saints doing more or less well under colonial rule in Senegal and Northern Nigeria see C. Coulon, *Le Marabout et le prince* (Paris: Pedone, 1981), 145–207; D. M. Last, 'Aspects of Administration and Dissent in Hausaland 1800–1968', *Africa*, 40/4 (1970); Cruise O'Brien, *Saints and Politicians*, pp. 85–111; P. B. Clarke, *West Africa and Islam* (London: Edward Arnold, 1982), 184–234.

Racial confrontation is often a significant ingredient in the charismatic appeal, although it could take local character in various ways. Thus Mourides or Hamallists might present their enemy as being the European colonialist persecutor; but the Qādiriyya of Bagamoyo in colonial Tanganyika defined itself above all as independent from the Arabs and Swahili of the coast. These latter Qādiris were the beneficiaries of colonial modernization, and indeed of the British colonial administrative policy of indirect rule, as of course were the Qādiris of northern Nigeria. And in the case of the Yoruba Mahdists studied by Peter Clarke in Chapter 8 the charismatic community defines itself as independent of both Europeans and Arabs: 'The Whites and the Reds had their own prophet, the Blacks must have their own too . . . are we not also made by God?' This peaceable Mahdist community, which awaited the end of British colonial rule together with the millennium, had from its outset (1942) the support of the aspirant politicians of the day; here nationalism was moving from the sacred to the secular.

The establishment of the Islamic brotherhood as a political institution, actively involved in the organization of worldly as well as spiritual affairs, is seen as the major accomplishment of the African Muslim saints of the period from 1750 to 1920. And charisma was the decisive ingredient in this transformation of the *ṭarīqa*: the miraculously validated personal authority of the saint was to appeal to a mass of new recruits, making possible what Murray Last has termed a 'populist' Islam. Lay members, aspirants or *murīd*s, were thus to be actively recruited for the first time, and then taught the value of a rigorous discipline (what Professor J. S. Trimingham has termed 'a spiritual slavery').[28] The disciple or *murīd* was to obey his *shaikh* in every way; he was to resemble, in the expressive Sufi phrase, 'a corpse in the hands of the embalmer'. And upon this basic relationship, between lay disciples and spiritual guides, was built an elaborate hierarchy of authority in the Islamic brotherhood.

The *ṭarīqa* in previous times, before the mid-eighteenth century, had something of the character of a secret society, with membership reserved for a learned élite; it might be seen as a gentlemen's club. A Muslim scholar and gentleman then usually belonged to more than one such club, and within each *ṭarīqa* he

[28] J. S. Trimingham, *The Sufi Orders in Islam* (Oxford: Clarendon Press, 1971), 103–4.

might seek the guidance and patronage of more than one spiritual mentor. The charismatic Muslim saint was thus to innovate in insisting on an exclusive allegiance to one *ṭarīqa*, his own, and in seeking to recruit lay members as extensively as possible. The notion of exclusive membership (of a single *ṭarīqa*) reflects the more demanding authority of the saint and is perhaps a necessary part in the new organization of the brotherhood. While 'membership' in several brotherhoods had been possible for a single individual, one might have regarded oneself as fully bound to none in particular. The charismatic saint sought then to convert the *ṭarīqa* into something that was new in its sub-Saharan setting:[29] a corporate Muslim organization with a disciplined mass membership.

The decisive organizational shift that was involved here is demonstrated by Louis Brenner in Chapter 2, juxtaposing the cases of Sidi al-Mukhtar al-Kunti (1729–1811) and the Shehu ʿUthman b. Fudi (1754–1817). The Shehu, who was affiliated to several religious orders, saw *ṭarīqa* membership as 'a privilege reserved for the *ʿulamāʾ*'; the common people would receive their necessary instruction from the learned *ʿulamāʾ*, but they would not be members of the club. Sidi al-Mukhtar, on the other hand, was concerned to use the *ṭarīqa* as a political instrument. His own brotherhood, the Qādiriyya-Mukhtāriyya, centralized around his own sainthood, could become an effective instrument once he had established its claims to spiritual superiority. A mass membership would automatically follow, the disciple-members owing an absolute obedience to their spiritual leaders. The brotherhood could thus become involved in the organization of social action in this world as well as providing the path to paradise in the next.

The activist Muslim brotherhood, stressing the devotional value of practical activities in commerce or in farming, was thus to be a prevalent pattern in the nineteenth-century 'revival' of the *ṭarīqa*.[30] The new disciples, recruited under charismatic

[29] The organization of the Sufi brotherhood in Islam, and the trend to *ṭāʾifa*, are charted by J. S. Trimingham from the fifteenth century; see *The Sufi Orders in Islam*.

[30] On the nineteenth-century revival or creation of brotherhoods see B. G. Martin, *Muslim Brotherhoods in Nineteenth-Century Africa* (Cambridge: Cambridge University Press, 1975); and for a critical reappraisal of the significance of brotherhoods in West African Islam see J. L. Triaud, 'Le Thème confrèrique en Afrique de l'Ouest', in A. Popovic and G. Veinstein (eds.), *Les Ordres mystiques dans l'Islam: Cheminements et situation actuelle* (Paris: EHESS, 1986).

auspices, could then be used as a labour force (for example by Shaikh Sidiyya al-Kabir to the north of the Senegal valley) or indeed as an army (notably so by al-Hajj ʿUmar Tal in his holy war from Senegambia to the Niger bend).[31] Islamic renewal under the guidance of such saintly leaders took on the character of a political movement, tightly disciplined and given to maximum mass recruitment. This period in Sufi history roughly corresponds to the 'third stage' in the evolution of the Sufi order, as presented by Professor Trimingham: this is the stage of *ṭāʾifa*, or of organization, when Sufism becomes a popular movement.[32] The new disciple 'surrenders' to a person with the gift of grace in this last stage of Sufi development, the time of 'spiritual slavery'.

A pluralism of mystical allegiance could thus no longer be tolerated as Sufism entered the public domain in the nineteenth century. Before this time it had been the case that Sufism tended in the direction of esoteric practice as well as aristocratic recruitment: 'participation in the *ṭarīqa* could take the form of a personal religious discipline under the guidance of one or several spiritual masters, as well as a wide-ranging quest for the acquisition of efficacious prayers and secrets' (Louis Brenner, in Chapter 2). Many Sufis in this earlier period had kept their brotherhood affiliations secret, or at least private. The contribution of charismatic leadership, then, was to bring Islamic practice into the popular age, by the public promotion of exclusive claims on behalf of one particular Sufi leader, and one particular brotherhood growing down from that leader. The Sufi brotherhood thus became a coherent organization for the first time under Sidi al-Mukhtar al-Kunti: it took his charisma to create the brotherhood, south of the Sahara.

It is suggested here that the saint may promote the idea of his own sainthood (*walāya*), his own spiritual superiority, and recruit his own mass following on this basis. Thus the adept careerist Sidi al-Mukhtar 'went to great lengths to distinguish his own *ṭarīqa* (the Qādiriyya-Mukhtāriyya) from others and to proclaim its superior qualities with the most efficacious prayers

[31] On ʿUmar Tal see Robinson, *The Holy War of Umar Tal*; also F. Dumont, *L'Anti-Sultan ou al-Hajj Omar Tal du Fouta, combattant de la foi* (Dakar: Nouvelles Éditions Africaines, 1974); J. R. Willis, *The Passion of Hajj Umar* (London: Frank Cass, forthcoming).

[32] On *ṭāʾifas* see Trimingham, *The Sufi Orders in Islam*, pp. 67–104.

to be recited in that era' (Louis Brenner, in Chapter 2). The North African founder of the Tijāniyya brotherhood, Ahmad al-Tijāni (1737–1815), went a stage further in claiming to be the 'intercessor of the age', the superiority of his brotherhood having been endorsed by the Prophet Muhammad (in a vision, when awake). The Tijāniyya brotherhood, developing in sub-Saharan Africa from the time of al-Hajj 'Umar Tal (1796–1864), was to stress a strict exclusivism of allegiance: no other brotherhood's prayer formulae were to be recited by the Tijāni disciple, for whom there could be no other saint but Shaikh Ahmad and his successors.[33] The Tijāniyya was then to develop during the colonial period as a characteristically urban brotherhood, as in Senegambia, where it recruited widely among casted persons and ex-slaves who sought refuge in town from rural oppression.[34] The branch of Tijāniyya led by Ibrhim Niass of Senegal, the 'Community of Grace', was to become probably the largest single Muslim organization in Africa; this branch, particularly successful in Nigeria, has been styled 'reformed Tijāniyya', pointing to its special concern for the teaching of the Arabic language and the Islamic sciences, as well as its pan-Islamic orientation.[35] But the Tijāniyya is not alone among Islamic brotherhoods in adapting to an urban environment; another example is provided by the Hausa traders studied by Professor Abner Cohen,[36] and there is a comparable modern adaptation of the *ṭarīqa*, among the Mouride brotherhood of Senegal (see D. B. Cruise O'Brien in Chapter 7).

When membership of the *ṭarīqa* had been focused upon the saintly person, then the disciple's duty of absolute spiritual

[33] See J. Abun Nasr, *The Tijaniyya: A Sufi Order in the Modern World* (London: Oxford University Press, 1960), 15–26. On the origins and early growth of the African Tijāniyya see Robinson, *The Holy War of Umar Tal.*

[34] L. Colvin, 'The Shaykh's Men: Religion and Power in Senegambian Islam', in Fisher and Levtzion (eds.), *Rural and Urban Islam in West Africa*, pp. 70–1.

[35] Hiskett, *The Development of Islam in West Africa*, pp. 286–8. On recent developments in the West African Tijāniyya, see also id., 'The "Community of Grace" and its Opponents the Rejecters', *African Language Review*, 17 (1980), 99–140; B. G. Martin, 'Les Tidianis et leurs adversaires: Développements récents de l'Islam au Ghana et au Togo', in Popovic and Veinstein (eds.), *Les Ordres mystiques dans l'Islam.*

[36] A. Cohen, *Custom and Politics in Urban Africa: A Study of Hausa Migrants in Yoruba Towns* (London: Routledge and Kegan Paul, 1969); also id., 'Cultural Strategies in the Organisation of Trading Diasporas, in C. Meillassoux (ed.), *The Development of Indigenous Trade and Markets in West Africa* (Oxford: Oxford University Press and International African Institute, 1971).

obedience could be turned to political and economic purposes, to become a force with which governments and opponents must reckon. A good example here, presented by Professor Charles Stewart, is that of Shaikh Sidiyya al-Kabir (1776–1868), who used his brotherhood (Qādiriyya) for the organization of long-distance trade, effectively excluding all commercial competition, with its own routes and wells in the semi-desert to the north of the Senegal river. This Mauritanian saint, who learnt the arts of sanctity from Sidi al-Mukhtar al-Kunti, and who helped to teach them to Amadu Bamba, was also an accomplished political mediator (something of a Sufi speciality), and even a king-maker among the warrior Hassani.[37]

The brotherhood, active in the management of the affairs of this world as well as in providing a vehicle for salvation in the next, is to be seen as the Sufi form of charismatic routinization. The charisma of the founding saint, the anchor which holds the brotherhood in place, is preserved after his death by rituals surrounding the sacred tomb. All members of the brotherhood should visit the tomb, where they will receive the saint's blessing (*baraka*). Such a pilgrimage is a reminder of the brotherhood's common purposes, variously evident otherwise in business or in politics. The maintenance of a charismatic authority is, however, highly problematic: the saint's successors, heirs to his miraculously based charisma, have the evanescent resource of the extraordinary turning to the routine: their authority is no longer that of the miraculous, and is extraordinary now only in degree. But the brotherhood, turning sacred memory to organizational advantage, has, in the opinion of Professor Trimingham, come as close to constituting a church as is possible within Islam, its hereditary clerics approximating to a priesthood.[38] The authority of the *shaikh*, be his charisma never so routinized, has usually been wide enough to make his spiritual disciples also into a political clientele—whose loyalty can be delivered to secular government (colonial, single-party, or presidential), always at a price.

The absolute obedience which was expected of the Sufi disciple attracted early interest from European colonial authority in Africa, and brotherhoods came to prosper under colonial

[37] Stewart, *Islam and Social Order in Mauritania*, pp. 145–7.
[38] Trimingham, *The Sufi Orders in Islam*, pp. 70–83.

rule in various territories (notably in Senegal, Mauritania, northern Nigeria, Tanganyika, Sudan, and Somaliland).[39] The French authorities in Algeria, France's first African colony, were by the mid-nineteenth century discovering the possible uses of Sufi brotherhoods in a variant of administrative 'indirect rule'[40] (following an unacknowledged Ottoman precedent). The Algerian model was then to be of great significance in the establishment of France's sub-Saharan administrations at the end of the nineteenth century. The colonial government, in tacitly recognizing the Islamic brotherhoods for reasons of administrative expediency, in order to collect the necessary impositions (in blood, sweat, and taxes) without undue disturbance of the peace, also helped to alter the basis of Sufi authority. The cohesion of reputedly loyal brotherhoods was reinforced, while the reputedly disloyal were pursued and chastised.[41] Chains of spiritual authority within the *ṭarīqa* were redefined in the language of political command, with an administrative parcelling of holy authority—the first tentative steps towards a bureaucratic routinization of charisma.

Europe thus has an important part in the stimulation of change in Muslim charismatic authority, but it is also a part to be understood as limited in scope. It is true that colonial government, British as well as French, bolstered some Sufi brotherhoods while using their leaders as administrative intermediaries. Funds and tickets to Mecca were directed towards Sufi notables in return for support of government programmes. But colonial manipulation had its limits, notably

[39] Trimingham, *Islam in the Sudan*; I. M. Lewis, *A Pastoral Democracy: A Study of Pastoralism and Politics among the Northern Somali of the Horn of Africa* (London: Oxford University Press and International African Institute, 1961); id., *A Modern History of Somaliland* (London: Weidenfeld and Nicolson, 1965). For an overview see Clarke, *West Africa and Islam*; also Nimtz, *Islam and Politics in East Africa*.

[40] See Harrison, *France and Islam in West Africa*, where it is argued that French policy in the 1920s and 1930s saw 'an unprecedented degree of French intervention in the internal affairs of the Sufi orders in order to maintain a strong and unified command' (p. 000). French confidence in their ability to intervene was based on the administrative research organized by Paul Marty from 1906 onwards. See also Triaud, 'Le Thème confrèrique', where it is argued that the colonial study of Muslim brotherhoods is riddled with misunderstanding, and that the post-colonial study of the same subject inclines to *Sénégalocentrisme* or even *Mouridocentrisme*.

[41] For an interesting case of colonial persecution see L. Sanneh, 'Tcherno Aliou, the Walī of Goumba: Islam, Colonialism, and the Rural Factor in Futa Jallon 1867–1912', in Fisher and Levtzion (eds.), *Rural and Urban Islam in West Africa*.

in a deficiency of reliable information; there was also manipulation, probably much more effective, from the African Muslim side (consider for instance the role of the Waziri in Joyce Carey's African novels). Although the Muslim brotherhoods' leaders were willing in most cases to co-operate with colonial authority, on condition that they received administrative favour for themselves and for their disciples, this was no affair of the heart—for all the self-congratulatory delusions of European power. When the end of European rule became foreseeable, after the Second World War, many important brotherhoods gave their support to the likely heirs in the nationalist parties.[42] Muslim brotherhoods in such cases were very useful to the politicians, getting out the vote for the crucial elections of the 1950s—*le temps de la politique*, as they say in the Ivory Coast, the time when political destinies were decided at the ballot-box. The winners were the Western-educated, victorious upon occasion through the unlikely democratic agency of the hierarchies of the *ṭarīqa*; so the direct exercise of power was thenceforward to depend upon a formal education which had been little favoured by the Sufi notables, as power became increasingly concentrated in the cities.

Then there is the politics of the supernatural: Sufi power was by no means restricted to electoral mediation on behalf of an aspirant political élite. The domain of magic, too little studied in the academic literature on African politics, should at least be mentioned here. The Sufi was generally respected and sometimes feared for his magical prowess; he sold his power (*baraka*) to the anxious politician, in the form of charms and protective formulae. Many African political leaders (e.g. Kwame Nkrumah, who sought it from Ibrahim Niass of the Tijāniyya) are known to have sought the support of such Sufi power. This under-studied area of African politics is suggestively covered, in relation to the subject of competitive therapy, by Murray Last's contribution in Chapter 9, which refers to 'the apparently expanding industry of prayer and Islamic medicine evident

[42] See D. B. Cruise O'Brien, 'La Filière musulmane: Confrèries soufies et politique en Afrique Noire', in *Politique africaine*, 1/4 (1981); also V. Monteil, *L'Islam noir* (1st edn.; Paris: Éditions du Seuil, 1964); J. Cuoq, *Les Musulmans en Afrique* (Paris: Maisonneuve et Larose, 1975); I. M. Lewis (ed.), *Islam in Tropical Africa* (London: Oxford University Press, 1966).

today' in Kano. Islamic medicine (*ṭibb*), seen as highly effective in popular politics, is 'an essential ingredient in creating charisma'. Political instability in Nigeria since independence has therefore helped to promote a demand for the services of Muslim scholars, contributing to the development of a 'prayer economy' which plays an important part in the prosperity of modern Kano.

Other dimensions of Sufi politics are revealed in relation to the generation and distribution of wealth through the brotherhoods. As the leaders of the brotherhoods collect the disciples' pious offerings in cash or in kind, they can be seen as the preceptors of a spiritual taxation. These taxes, furthermore, have an exceptional status in that they are willingly (even enthusiastically) paid; the disciple is, as it were, delighted to be exploited, and he is proud of the wealth of his brotherhood and of his *shaikh*. But appearances of abject spiritual servitude should not deceive: the disciple does also expect a return on his tribute, and not only in the assurance of paradise. The brotherhood can provide the disciples with what amounts to the protection of an Islamic welfare state (albeit one of less than Scandinavian resources). For the Sufi lodge or *zāwiya* can discharge a wide range of valuable functions, acting as a court for the arbitration of disputes, a school for pious instruction, a bureaucracy for the administration of the brotherhood's affairs, a welfare office for the distribution of charity to the indigent, or an inn for the weary traveller. The generation of wealth through the brotherhood is at the same time assured not only by the collection of its 'spiritual taxation', but also by the organization of agricultural labour or of commerce through the auspices of the *ṭarīqa*. The Sufi establishment, in directing these many affairs, acts as a government in miniature, and the disciple in paying his 'spiritual taxation' is more likely to see a return on his taxes here than from the secular taxation of the state. The charismatic legitimacy of the brotherhood may be declining into patrimonialism, but its legitimacy at a popular level remains real: willingly paid taxes can perhaps be taken as the true sign of a legitimate authority.

The position of the Islamic brotherhoods is none the less today under serious threat. Some threats are chronic, such as the tendency to internal splitting which regularly saps the power of the *ṭarīqa* (although a few, such as the Senegalese Mourides or the Sudanese Mirghaniyya,[43] have maintained their unity and power). Other threats are produced by recent developments, such as the spread of popular literacy, which enables students and clerks to contest the Sufi leader's position as translator and interpreter of the sacred Arabic word; or the crisis in African agriculture, with the rural–urban population drift of the last two decades, which threatens the traditional power-base of the Sufi brotherhood (stimulating some to a creative redefinition of their socio-economic role). More field research relating to these recent developments is, however, urgently needed. Even if this Islamic institution truly is in general decline, precise investigation is still required; and in the more likely case that variations in the Sufi predicament are very broad from one state or brotherhood to another, these variations must be measured in comparative field research.

One generalization which might be tested is that it is government policy, with an intolerance of opposition in any form, which has been the greatest threat to the brotherhoods since African independence. The very fact that the Sufi brotherhood is capable of self-sufficiency (at least in ideal terms) may have led it into government disfavour. African governments have often preferred to promote the cause of national Muslim associations, directed by notaries who may be more amenable than the Sufi notables. Patronage expenditure can then be distributed, resources from the wider Muslim world going to selected local recipients, and state control over the Muslim community can thus be consolidated. This development has been accompanied by the progress of an Islamic bureaucracy, in an alliance of state power with the proponents of a 'modernized' or reformed Islam.[44] Reformists of the Wahhabiyya movement have become increasingly influencial in sub-Saharan Africa over the past twenty or thirty years; theirs is anti-Sufi Islam in the style of Saudi Arabia, an austere devotional style in which there is place for no other Islamic

[43] Trimingham, *Islam in the Sudan*; also J. O. Voll, 'A History of the Khatmiyyah Tariqah in the Sudan', Ph.D. thesis (Harvard University, 1969).

[44] See J. L. Triaud, 'L'Islam et l'État en république au Niger', *Le Mois en Afrique*, 192–3 (Nov.–Dec. 1982), 194–5 (Feb.–Mar. 1983).

charisma than that of the Prophet Muhammad.[45] The Sufi predicament with regard to government policy may, in the light of further field research, emerge as in fact quite widely various. In several of the states formerly under French rule centralization and concentration of power have proceeded at the expense of Sufi power: brotherhoods are almost banned in Niger, and the Tijāniyya is pursued by police action in Togo. But in Senegal, historically the 'paradise of brotherhoods',[46] the *ṭarīqa* continues to thrive. And in the states formerly under British rule a pattern of governmental centralization is much less easy to discern.

Political circumstances in Africa would seem propitious for the production of charismatic leadership, if a profound social crisis be the requisite seed-bed for charisma. The political institutions of the state are commonly under severe stress, as the agrarian economy is threatened by drought and over-taxation and the social fabric strained by mass urbanization. To see the problems in Islamic terms, there are many unjust rulers and much unbelief. When may one then expect the new generation of charismatic leadership to emerge in African Islam? Islamic charisma may perhaps be produced from the social base, as apparently in the case of the Maitatsine rebellion of 1980 in Kano, arising from the ranks of the recently urbanized shanty-towners.[47] Or it may be produced from the élite, with reformists such as Ahmed Khalifa Niasse, who declared an (abortive) Islamic republic in Senegal in 1979.

The failure of such recent Islamic insurrection in Africa does not of course foreclose future possibilities. The Muslim community south of the Sahara continues to grow rapidly, and it is in varying degrees ready to respond to original, challenging leadership. Such Muslim charisma may still emerge from the Sufi communities, which produced so many extraordinary

[45] See L. Kaba, *The Wahhabiyya: Islamic Reform and Politics in West Africa* (Evanston: Northwestern University Press, 1974); also Hiskett, 'The "Community of Grace" and its Opponents the Rejecters', *African Language Studies*, 17 (1980), 99–140; J. L. Amselle, 'Le Wahhabisme à Bamako', *Canadian Journal of African Studies*, 19/2 (1985); J. L. Triaud, 'Biographie d'un propagandiste Wahhabi originaire du Mali: Abd al-Rahman b. Yusuf al-Ifriqi al-Fatawi (1908–1957)', in O. Carré and P. Dumont (eds.), *Les Radicalismes islamiques* (Paris: l'Harmattan, 1985).

[46] G. Nicolas, *Dynamique de l'Islam au sud du Sahara* (Paris: Presses Orientalistes de France, 1981), 189.

[47] See A. Christelow, 'The Yan Tatsine Disturbances in Kano: A Search for Perspective', *The Muslim World*, 75 (1985).

people in the past. (Alternatively, the Sufis may opt for discretion: having gone public in the eighteenth century with Sidi al-Mukhtar al-Kunti, the brotherhoods may now, as in the Soviet Union, opt for withdrawal and clandestinity in the face of state power.[48]) It may also be that it is rather in Mahdism, the permanent revolution in Islam, that one should seek a new charismatic leader in Africa. One can perhaps discern in urban Muslim Africa the outlines of a charismatic constituency, awaiting the appropriate leader, although it would be impossible to specify in advance a time or a place. But then we have been dealing with the extraordinary, and the extraordinary is by definition peculiarly difficult to predict.

These studies, however, make no claim to foretell the future; their purpose is rather to interpret the past and present of African Islam. The discipline and intellectual effort involved in the emergence of saintly or prophetic leadership are seen as a variant of charisma, which has room for a substantial element of tactical judgement on the part of the leader in building his charismatic constituency. The aspirant Muslim leader must think through his campaign with pertinacity: lengthy studies, arduous retreat, then the afflux of pupil-disciples and the possible miracle—the Islamic equivalent of a presidential election campaign. And it is around a leader of this charismatic type that the brotherhood is built, a highly authoritarian mass-based institution in an African political environment remarkable for its uninstitutional character. Although the *ṭarīqa* may now be under threat, the ingredients for a new charismatic response do seem to be present. Africa certainly does not want for a crisis, aspirant Muslim leaders of application and quality are present, and a miracle is perhaps to be expected.

[48] For an instance of apparently successful Sufi withdrawal see A. Bennigsen and C. Lemercier Quelquejay, *Les Musulmans oubliés: L'Islam en Union Soviétique* (Paris: Maspéro, 1981).

## 2

# Concepts of *Ṭarīqa* in West Africa
# The case of the Qādiriyya

LOUIS BRENNER

MUSLIM scholars in West Africa often described the requirements of religious practice in terms of three theological concepts: (1) *Islām*, (2) *īmān*, and (3) *iḥsān*. Shaikh ʿUthman b. Fudi organized his book *Kitāb ʿulūm al-muʿāmala* (Book of the Sciences of Behaviour)[1] according to these designations:

1. *Islām* in this context denoted an individual's personal submission to God's law; Shaikh ʿUthman equated this submission with the science of *fiqh*, which he described as the outward requirements of religion. In this section of his book he discussed the rules for ablution, prayer, and so forth.
2. *Īmān* he equated with *tawḥīd* or theology, the study of the unity of God; here he discussed the attributes of God and of the prophets.
3. *Iḥsān*, the Shaikh said, was based on the science of *taṣawwuf*, which concerns the inward aspects of religion—the acquisition of praiseworthy qualities and the elimination of blameworthy ones.

*Taṣawwuf*, according to Shaikh ʿUthman, aided the Muslim in purifying the heart from the whisperings of Satan and from such moral disabilities as conceit, false hope, anger, and envy. *Fiqh*, *tawḥīd*, and *taṣawwuf* are all sciencies, or bodies of knowledge, which in Shaikh ʿUthman's time formed part of every religious studies curriculum. But these fields of study also differ from one another in what they demand by way of mastery. The *fiqh* to which Shaikh ʿUthman refers under the category of *Islām*, primarily ablution and prayer, can be learnt by mere imitation. The theological concepts of *tawḥīd* require considerably more intellectual application, even if they are simply committed to memory without any very deep understanding. And *taṣawwuf*,

[1] ʿUthman b. Fudi, *Kitāb ʿulūm al-muʿāmala*, trans. from the Arabic by ʿĀ'isha ʿAbd ar-Rahman al-Tarjumana, in *Handbook of Islam* (Norwich: Diwan Press, 1978).

which must be understood as a science of inner transformation, requires an even more sophisticated level of personal application. Study and reading, and even imitation of a teacher or *shaikh*, might encourage and support one's search for the development of inner qualities, but they cannot alone produce them. The methods associated with the religious science of *taṣawwuf* were developed and transmitted in the formal context of a *ṭarīqa*, under the direction of a spiritual guide who authorized one to recite certain prayers and directed one in specific exercises which were intended to aid one's inner development.

The word *ṭarīqa* in Arabic literally means method or manner of doing something. For Sufis the word bears the connotation of mystical way or path, or of religious rule. The word *ṭarīqa* is also often translated into English as 'order' or 'brotherhood'. This rendering is by no means alien to concepts of *ṭarīqa* held by the Sufis; they often refer to themselves as *al-ikhwān* (the brethren) or *al-qaum* (the folk). So they see themselves as an identifiable, corporate group. However, the nature of this corporateness, and the extent to which it is employed for social and political ends, varies considerably among Sufis. And these variations depend less on the particular doctrine or religious rule of the *ṭarīqa* than upon the intentions of a given leadership in the context of a given set of social, political, and religious conditions. As religious rules, the *ṭuruq* (plural of *ṭarīqa*) are very similar to one another, even if doctrinal disputes among them at times resulted in violent confrontations.

The *ṭuruq* institutionalized *taṣawwuf* and provided a structure and method for the pursuit of one's inner religious development which went well beyond the concept of *iḥsān* as exemplary behaviour. The aim of the *ṭuruq* was to bring the acolyte (*murīd*) into a closer relationship with God, ultimately to achieve union with him. As a religious rule, the *ṭarīq* included a set of prayers (the various *dhikr*, *wird*, *ḥizb*, etc.), spiritual exercises (such as *khalwa*), and disciplinary injunctions which define the relationship between the *murīd* and his *shaikh* (spiritual guide). Initiation (*talqīn*) into a *ṭarīqa* was effected by an agreement to submit to its disciplinary injunctions and by the granting of formal permission to recite its *wird*. Initially, a new *murīd* was allowed to recite only the *wird* or litany of the *ṭarīqa*; with time he would normally have been authorized to recite additional prayers. None of these prayers could be efficaciously recited without formal authorization by a recognized superior. In West Africa it was not uncommon for Sufis to collect many such authorizations, both from their own *shaikh* and from other prominent and respected members of the *ṭuruq*.

The collection of authorized prayers (and of 'secrets') reflects several significant aspects of the *ṭuruq* as they have existed in West Africa. The prayer itself, of course, is believed to be possessed of spiritual power and therefore to provide inherent benefits. But this power was dependent firstly on the origins of the prayer, which was normally composed by a notable Sufi *shaikh*, or perhaps transmitted to him in a vision by the Prophet or an earlier deceased *shaikh*; and secondly on the quality of the *silsila* (chain of transmission) through which it had been transmitted to its present possessor. The conscientious and pious *murīd* therefore sought not only to collect such prayers, but to do so from prominent and respected Sufis whose own authorizations were transmitted through the most prestigious *silsila*. The renewal of authorizations for prayers previously acquired, known as *tajdīd*, was therefore not unusual.

Participation in the *ṭuruq* could therefore take the form of a personal religious discipline under the guidance of one or several spiritual masters, as well as a wider-ranging quest for the acquisition of efficacious prayers and 'secrets'. In such conditions the concerns and interests of a given individual might take precedence over the demands of the *ṭarīqa* as a corporate entity. It may be that before the eighteenth and nineteenth centuries the *ṭuruq* existed in West Africa in this more diffuse form; it is probable that before this period many Sufis kept their affiliations secret, or at least private. In the eighteenth century with Sidi al-Mukhtar al-Kunti, and in the nineteenth century with al-Hajj ʿUmar al-Futi, the *ṭuruq* began increasingly to emphasize their distinctive identities and corporateness; concurrently they became public institutions.

The emergence of the *ṭuruq* in West Africa as visible, coherent organizations was not an accident but the result of conscious decisions by Sufi leaders who saw in them a potential not only for religious change, but also for social and political transformation. Both Sidi al-Mukhtar and Shaikh ʿUthman presided over such transformations, but they did not both perceive or employ the potential of the *ṭuruq* in the same way. However, they did both derive much of their authority, both religious and political, from their ascribed *walāya*. *Walāya* is often translated as 'sainthood', but a more appropriate connotation of the word is 'nearness to God'. The *walī* is close to God, a friend of God; he represents the ultimate attainment of spiritual quest in the *ṭarīqa*. In Sufi theory, the *walī* receives his elevated state as a gift from God; his piety and religious exercises are only preparations to receive this gift. His visions and *karāmāt* (wonders or miracles) are not of his own making, but occur through or around him by divine agency. In social practice, however, the *walī* is a source not only of *baraka*, but also of power: the power to bless, to heal, to judge, to foresee, to transform, and also, through his curse, to bring down the punishment of God. The *walī* has direct access to divine intervention; he has been elevated to a position above social convention; potentially he is the supreme innovator. Perhaps most importantly, from the perspective of social and political transformation, the *awliyā'* (plural of *walī*) were often seen, and portrayed themselves, as opposed to established political authority.

This chapter will compare the conceptions of *ṭarīqa* held by Sidi al-Mukhtar al-Kunti and Shaikh 'Uthman b. Fudi. The influence of these men has been extensive in West Africa; they both directed movements of considerable social, political, and religious significance. Both were also affiliated to the Qādiriyya *ṭarīqa* and were thought by many of their contemporaries to have attained the status of *walī*. But their understanding of *ṭarīqa* differed, not in their devotion to it as a religious rule, but in their perceptions of it as an effective model for social and political organization.

## 1. SIDI AL-MUKHTAR AL-KUNTI (1729–1811)

This religious leader clearly perceived the social and political, as well as religious, potential of the *ṭarīqa* as a corporate entity. The analysis which follows is based largely on the research of A. A. Batran, who has studied in depth the extensive writings of the

Kunta.[2] From these sources Batran has been able to reconstruct a convincing biography of Sidi al-Mukhtar, and an account of the role which *taṣawwuf* and *ṭarīqa* played in his rise to eminence and influence. There are problems of historiography, of course: the major biography of Sidi al-Mukhtar is hagiographic, and naturally idealizes his person and exaggerates his miraculous powers.[3] But the vast corpus of Sidi al-Mukhtar's own writings contains ample evidence of his own personal conviction of his *walāya*, and of a determined effort on his part to employ the powers and influence which this *walāya* implied to mould a coherent and integrated Muslim community around himself. At the centre of this community was the *ṭarīqa* of which Sidi al-Mukhtar was the undisputed leader and which came to be known as the *Qādiriyya-Mukhtāriyya*. Here, an attempt will be made to demonstrate how Sidi al-Mukhtar's personal vision of *ṭarīqa* imparted to the *Qādiriyya-Mukhtāriyya* its own unique character.

## *The milieu*

It must be emphasized from the outset that the star of Sidi al-Mukhtar's *walāya* did not ascend unimpeded into the heavens. The hagiographies might give this impression, but Batran

[2] The best source for Sidi al-Mukhtar's career is A. A. Batran, 'Sidi al-Mukhtar al-Kunti and the Recrudescence of Islam in the Western Sahara and the Middle Niger', unpublished Ph.D. thesis (University of Birmingham, 1971), hereafter referred to as Thesis. Portions of the thesis have been published in articles: see Batran, 'An Introductory Note on the Impact of Sidi al-Mukhtar al-Kunti (1729–1811) on West African Islam in the 18th and 19th Centuries', *Journal of the Historical Society of Nigeria*, 6/4 (June 1973), 347–52; id., 'The Kunta, Sidi al-Mukhtar al-Kunti, and the Office of *Shaykh al-Ṭarīqa*' al-Qādiriyya', in J. R. Willis (ed.), *Studies in West African Islamic History*, i. *The Cultivators of Islam* (London: Frank Cass, 1979), 113–46; id., 'The *Qādiriyya-Mukhtāriyya* Brotherhood in West Africa: The Concept of *taṣawwuf* in the Writings of Sidi al-Mukhtar al-Kunti (1729–1811)', *Transafrican Journal of History*, 4/1–2 (1974), 41–70. For the general background to this period see J. R. Willis, 'The Western Sudan from the Moroccan Invasion (1591) to the Death of al-Mukhtar al-Kunti (1811)', in J. F. A. Ajayi and M. Crowder (eds.), *History of West Africa*, (London: Longman, 1985), i. 531–76.

[3] Muhammad b. al-Mukhtar al-Kunti, *al-Ṭarā'if wa'l-talā'id bi karāmāt al-Shaykhayn al-wālida wa'l-wālid*. This work is unpublished, although manuscript copies are easily available, e.g. in the Bibliothèque Nationale, Paris, Manuscrits Orientaux, Arabe 5334, fos. 1–555, and Arabe 5511, fos. $12^a$–$111^b$. Translated extracts can be consulted in I. Hamet, 'Littérature arabe saharienne', *Revue du monde musulman*, 12 (1910), 196–213, 380–405.

makes it clear that al-Mukhtar's religious community was forged in the midst of considerable economic, political, and religious competition. Each stage in the development of this community was marked by conscious decisions based on a particular view of contemporary problems and issues, and of Islam's possible responses to them. Unfortunately, a co-ordinated chronology of specific issues and the development of al-Mukhtar's thought has yet to be compiled, but the general pattern is clear enough. Sidi al-Mukhtar felt that he lived in an age of advanced decadence which had begun some centuries earlier. This decadence was not limited to the southern Sahara where he lived, but extended to all the lands of Islam. He also came to feel that he had been chosen by God as the *mujaddid* or 'renewer' of Islam in his era,[4] who was destined to reverse this trend, not by extending Islam through force of *jihād*, nor by imposing it upon others, but by establishing a religious community which could serve as a refuge for those Muslims who wished to renew their own faith. The strategy was one of appeal rather than demand, of example rather than condemnation. This community was organized and articulated through personal and group relationships: firstly through the kinship networks of the Kunta, further afield through the personal religious bonds among *mashā'ikh* and *murīd* within the *ṭarīqa*, and finally through an extended network of *zawāyā* (Sufi religious centres). Nothing in these arrangements or concepts was original with al-Mukhtar, but in the context of time and place this new organization was ambitious and innovative. He directly challenged existing *zawāyā*[5] organizations such as the Kel Intasar and the Kel al-Suq, and claimed that in his time the only 'pure' Islam was found in his own community.

Sidi al-Mukhtar insisted that all members of his *ṭarīqa* should earn their own living and not depend upon the charity of others. Of course, the Kunta had for generations been active in trans-Saharan commerce, so this idea too was not original. But he placed great emphasis upon the accumulation of wealth, and insisted that there was a clear link between economic success

[4] Batran, Thesis, pp. 164 ff.

[5] The word *zawāyā* has two meanings in the context of Saharan Islam: *zawāyā* is a social designation denoting Islamic clerical groups in contradistinction to the so-called warrior lineages; it is also the plural of *zāwiya*, a Sufi religious cell or study-centre.

and religious piety. Wealth, he said, was an indication of one's *jah* (dignity and status). His son, Sidi Muhammad, drew the following portrait of the ideal Kunta career:

> When the Kunta youth reached maturity their ultimate desire would be to travel in search of knowledge. After acquiring a substantial grounding in the religious and mystical sciences they would concentrate on accumulating wealth. Then they would seek to marry into the most noble families. Each one of them would become distinguished for his knowledge, piety and wealth.[6]

Sidi al-Mukhtar aggressively developed Kunta commercial enterprise, which naturally brought him into competition and conflict with many groups—competing merchants, exploitative desert raiders, and political authorities who sought to levy excessive taxes. He employed every means at his disposal to overcome his competitors and adversaries, including diplomacy, sharp business practice, playing the role of neutral mediator in disputes, and employing the awesome power of his curse (*da'wa bi- 'l-sharr*).[7] Failing all else, his people also took up arms.

The religious and economic networks developed by Sidi al-Mukhtar were directed virtually exclusively, it would seem, by the Kunta Awlad al-Wafi. Batran also observes that during al-Mukhtar's lifetime the *mashā'ikh* of the Qādiriyya Mukhtāriyya were appointed exclusively from the Kunta. Sidi al-Mukhtar's Islamic community was therefore centred on a Kunta ethnic identity, its layout organized according to the demands of an extensive commercial network, its hierarchy after the fashion of a Sufi *ṭarīqa*. His scholarly reputation also attracted numerous students and scholars to him. This was an expansive Muslim community, not through forceful means, but by voluntary association. However, the influence and authority of its leader extended far beyond the boundaries of those formally associated with him. According to Batran, the essential ingredient in the extraordinary success of al-Mukhtar's initiatives was the widespread acceptance of his *walāya*.[8]

[6] Batran, Thesis, p. 235, quoting from Muhammad b. al Mukhtar, *Ṭarā'if*.
[7] Ibid. 275 ff.
[8] Ibid. 134, 147.

## Walāya

*Walāya* may well be a social phenomenon, in the sense that without followers there can be no *walī*; but it is equally a personal phenomenon. *Walāya* does not occur by spontaneous generation; the 'germ' of *walāya*[9] must also be present in the person, and Sidi al-Mukhtar certainly seems to have possessed it. Even taking into account the exaggerations of the hagiographers, it seems evident that al-Mukhtar was an exceptionally gifted person. The basis of his spiritual and intellectual development is attributed to the four years he spent under the guidance of Sidi ʿAli b. Najib, with whom he began to study in 1743 (that is, when he was fourteen years of age). It was from Sidi ʿAli that al-Mukhtar received the Qādiri *wird*, and to him that he looked as his spiritual master.[10] After Sidi ʿAli's death in 1747 al-Mukhtar went to Walata, where he became the guardian of the tomb of his illustrious ancestor Shaikh Ahmad Bakka'i. In this position he made contact with numerous pilgrims who visited the tomb, and this constituted the first steps in building up the relationships which would later cement his religious community. Subsequently (1748–53) he spent a period in Taqant, where he began to dispense the Qādiri *wird* and to act successfully as a mediator of disputes. He was still only in his early twenties, and many of the local *ʿulamā'* opposed him or tried to resist his growing influence, claiming he was too young.[11] The biographies assert that Sidi al-Mukhtar overwhelmed his detractors with the breadth of his knowledge. This is probably true; he was undoubtedly extraordinarily brilliant. He was also undoubtedly very pious. Thus even in his youth he manifested clear signs of the gift of *walāya*.

The early appearance of exceptional qualities is a common pattern among the attributes of the *awliyā'*. Sidi al-Mukhtar's depth of learning, the power of his intellect, and the force of his pious example (perceived as the fruits of his spiritual training) all contributed to the germination of his *walāya*. These were the visible signs; there were also, of course, the visions and the *karāmāt*, the esoteric signs of divine favour. His earliest vision

[9] This concept was introduced into our discussions about charisma by M. Jean Séguy.
[10] Batran, Thesis, p. 125.
[11] Ibid. 106–8.

occurred in Walata (probably at the age of nineteen). Subsequently he received visionary evidence that he was the *mujaddid* or renewer of Islam in his age, when the Prophet appeared to him and said: 'You are the true messenger to renovate the *sunna*; be patient.'[12] Of course, the biographies attribute to him numerous *karāmāt*. All of these factors, taken together with the tangible success of his political and economic endeavours, provided the circumstances from which Sidi al-Mukhtar's *walāya* emerged, and by which it was confirmed and sustained.

In later years he would make many additional claims for himself, such as the right to *ijtihād* in legal matters (although he seems to have carefully avoided making legal judgments in specific disputes) and the possession of *ḥifẓ*, divine protection from sin and error.[13] And we have already mentioned that he threatened and employed the power of his curse against his enemies. But among all the practical implications of his *walāyā*, probably the most significant is to be found in the authority which it imparted to the organization of his *ṭarīqa*, because the Qādiriyya-Mukhtāriyya was at the heart of the social and political transformation which Sidi al-Mukhtar brought about.

## Ṭarīqa

Sidi al-Mukhtar al-Kunti either assumed or was endowed with the title *shaikh al-ṭarīqa al-Qādiriyya*. Batran is uncertain about the actual procedure through which this title was granted, but the predominantly diffuse nature of *ṭarīqa* organization in the eighteenth century in this region would suggest that al-Mukhtar claimed the title for himself, perhaps authorized by a vision to do so. It is certainly not the kind of title which could be granted by another living person; Sidi al-Mukhtar asserted that the rank of this title was equal in dignity to that of ʿAbd al-Qadir al-Jilani himself, with the consequence that he owed no personal allegiance to any other living person.

The hierarchy of authority as well as membership in the *ṭarīqa* was based upon two widely accepted models of relationship: blood descent and formal initiation. At the summit of this

[12] Ibid. 172, apparently quoted from a poem by Sidi al-Mukhtar.
[13] Ibid. 174 ff., 137.

hierarchy was Sidi al-Mukhtar himself, who as *shaikh al-ṭarīqa* derived his authority from the divine sources of *walāya*. Below him were the following rankings.[14]

(1) The *ahl al-Shaikh Sidi al-Mukhtar* were his sons, who were all addressed by the title *Sidi*.
(2) The *mashā'ikh* were selected exclusively from the Kunta, and were addressed as *Shaikh*.
(3) The *talāmīdh* were those who received the *wird* from Sidi al-Mukhtar or from the other *mashā'ikh*, and who were in turn authorized to transmit the *wird* and to found *zawāyā*; this seems to have been the highest rank attainable by non-Kunta in the Qādiriyya-Mukhtāriyya during al-Mukhtar's lifetime.
(4) The *murīdūn* were those who had received the *wird* from the *talāmīdh*, or who had not taken the *wird* but acknowledged the religious authority of Sidi al-Mukhtar.

The closer one's relationship to al-Mukhtar, the higher one rose in this hierarchy. However, blood took precedence over religious initiation. This procedure may well have been due to considerations imposed by the conditions of desert society and political relationships. In any case, the *ṭarīqa* was employed as a means to cement blood ties, and to complement and extend them.

The strength of the *ṭarīqa* rested firmly on the acceptance of Sidi al-Mukhtar's *walāya*. In addition to all the usual claims of visions and *karāmāt*, al-Mukhtar went to great lengths to distinguish his own *ṭarīqa* from others and to proclaim its superior qualities.[15] He asserted that his *dhikr* was the most effective of medicines then available 'for the human soul'. He also wrote about the extensive benefits in this world and the next which the *talāmīdh* of his *ṭarīqa* would receive if they persisted in their affiliations with him; however, he never went so far as the *Tijānis* would do in forbidding the *murīdūn* from associating with *mashā'ikh* from other *ṭuruq*. He also composed an extensive collection of prayers, known as the *awrād* and *aḥzāb* of the Qādiriyya-Mukhtāriyya.[16] These were divinely inspired, he

[14] Ibid. 156 ff.
[15] Ibid. 210 ff.
[16] Ibid. 223 ff.

said, to be transmitted through him to his contemporaries as the most efficacious prayers to be recited in that era. Other prayers were not forbidden, but those of the Qādiriyya-Mukhtāriyya were the best available at the time.

In keeping with political patterns of the southern Sahara, Sidi al-Mukhtar attempted to distinguish the extensive social and political organization which he had established from any association with secular or state structures. He did not impose his authority; people submitted to him by voluntary initiation into the *ṭarīqa*. He did not levy taxes; his followers brought him gifts, which he redistributed. He did not surround himself with pomp and panoply; he lived in simple, modest surroundings. He did not impose judgments; he mediated disputes. His followers did sometimes take up arms, but only in extreme circumstances, according to them; al-Mukhtar preferred to employ the power of his curse in order to discipline his enemies. The Qādiriyya-Mukhtāriyya provided a model of social and political organization which contrasted starkly with existing forms of secular authority in the region.

## 2. SHAIKH ʿUTHMAN B. FUDI (1754–1817)

In many ways, the life of Shaikh ʿUthman b. Fudi (1754–1817) followed a pattern similar to that of Sidi al-Mukhtar.[17] Shaikh ʿUthman also manifested exceptional qualities of intellect and self-assurance in his youth; he was already preaching in public and imploring others to the proper practice of *sunna* in his twenties. He too experienced visions which justified his proselytism, and he too claimed to be a *mujaddid* or rejuvenator of Islam in an age of religious decadence. His *walāya* was also widely accepted, and therefore played a central role in the

[17] There are many published works on the life of Shaikh ʿUthman and his *jihād*: M. Hiskett, *The Sword of Truth: The Life and Times of the Shehu Usuman dan Fodio* (New York: Oxford University Press, 1973); F. H. El-Masri, 'The Life of Shehu Usuman dan Fodio before the *jihād*', *Journal of the Historical Society of Nigeria*, 2 (1963–4), 435–48; ʿUthman b. Fudi, *Bayān wujūb al-hijra ʿalā ʾl-ʿibād*, ed. and trans. from the Arabic by F. H. El-Masri (Oxford: Oxford University Press, 1978), introduction. See also D. M. Last, *The Sokoto Caliphate* (London: Longman, 1967). There are also internal Arabic sources on the life of the Shaikh: ʿAbd Allah b. Muhammad Fudi, *Tazyīn al-Waraqāt*, ed. and trans. from the Arabic by M. Hiskett (Ibadan: Ibadan University Press, 1963); and Muhammad Bello, *Infāq al-maisūr fī taʾrīkh bilād al-takrūr*, ed. C. E. J. Whitting (London: Luzac, 1957).

extension of his reputation and the establishment of his authority. But there were also important differences between these two great men. Shaikh ʿUthman's religious community developed within the confines of an established state, not within the interstices of scattered and diffuse desert communties. Nor was he a merchant needing to respond to the commercial demands of an extended network of trade. The material resources of his community, and the material demands upon it, were therefore very different to those experienced by Sidi al-Mukhtar. These factors were undoubtedly crucial in determining the nature of Shaikh ʿUthman's religious community and the role which concepts of *ṭarīqa* would play in it.

## *The milieu*

Shaikh ʿUthman began his public career in a much more modest fashion than did Sidi al-Mukhtar. His basic concern was the education of Muslims in correct Islamic practice; he was a preacher with legalistic inclinations. It is doubtful whether in these early years he envisaged establishing a Muslim community apart from the one in which he was already living, or even whether he saw himself in conflict or competition with other Muslims. The earliest disputes which his activities aroused found him advocating a very tolerant and open position on the question of who was and who was not a Muslim.[18] He seemed to be seeking the widest possible definition of the Muslim community, with a view to educating all persons who considered themselves Muslims in the legal demands of the *sharīʿa*. His visions also began rather later in life than those of Sidi al-Mukhtar; the most politically significant of these occurred in 1794, at the age of forty, which suggests that his affirmation of any personal *walāya* was not relevant to the early stage of his career.

Of course, Shaikh ʿUthman did begin to attract a following around himself, which would form the embryo of a distinct

[18] This tolerant attitude is evident in the Shaikh's *Iḥyāʾ al-sunna*, completed in 1793; see I. A. B. Balogun, 'A Critical Edition of the *Iḥyāʾ al-sunna waʾl-ikhmād al-bidʿa* of ʿUthman b. Fudi, Popularly Known as Usumanu dan Fodio', unpublished Ph.D. thesis (School of Oriental and African Studies, University of London, 1967). A similar tolerance is present in a later, more polemical, work, *Naṣāʾiḥ al-ummat al-muḥammadiyya*, University of Ibadan, MS 82/94, undated but probably written between 1795 and 1803.

Muslim community when opposition to his activities began to take shape. Opposition came from some dissident or disagreeing scholars, but most significantly it came from the authorities of Gobir, the state in which the Shaikh lived most of his adult life. The separate identity of Shaikh ʿUthman's own *jamāʿa* or religious community developed in direct proportion to the growing official enmity of the state, which led ultimately to *hijra* and *jihād*. With the outbreak of open warfare and the prospect of establishing a state of their own, Shaikh ʿUthman and his fellow jihadists were required to view the Muslim community from a perspective very different from that which had been appropriate during the early years of preaching. Firstly, they adopted a much less tolerant position on the question of who was a Muslim;[19] legally this new interpretation relied heavily on the writings of Abd al-Karim al-Maghili,[20] the fifteenth-century Algerian jurist, but practically this issue was taken up in order to justify the necessity to invoke *jihād* against persons who themselves claimed to be Muslims. Shaikh ʿUthman went to great lengths to prove that his opponents were *not* Muslims. Secondly, there never seems to have been any doubt that the new Muslim *jamāʿa* should be organized according to classical Abbasid concepts of the state. The *ṭarīqa* model of organization contributed little or nothing to the formal structure of the new Sokoto Caliphate which emerged as a result of the success of the *jihād*.

## Walāya

Some authors have claimed that Shaikh ʿUthman's affiliations with the *ṭuruq* contributed significantly to his reformist inclinations and to the conduct of the *jihād*.[21] These views do not seem

[19] The first work in which the Shaikh adopted this new position was *Masāʾil muḥimma*, Bibliothèque de l'Institut de France, MS 2411 (180), written in 1803.

[20] For al-Maghili see J. O. Hunwick, *Sharīʿa in Songhay: The Replies of al-Maghili to the Questions of Askia al-Hajj Muhammad* (London: Oxford University Press with the British Academy, 1985).

[21] For example, B. G. Martin has said that the *jihād* movement 'depended on the ideology of the *Qādiriyya* brotherhood'; see *Muslim Brotherhoods in Nineteenth-Century Africa* (Cambridge: Cambridge University Press, 1976), p. 15. El-Masri has claimed that 'as leader of the Qādiri brotherhood, [the Shaikh] commanded blind obedience to himself'; see ʿUthman, *Bayān*, introduction, p. 5. Neither of these assertions is supported by available evidence.

justified in any specific social or political sense; there is no evidence that the Shaikh mobilized support or sought to organize his followers through the structure of a *ṭarīqa*. On the other hand, the intense demand for personal religious effort which was made by any serious pursuit of *taṣawwuf* was certainly important to him, as it was to Sidi al-Mukhtar. One might infer from the Shaikh's writings that one of his great hopes for his fellow Muslims, and perhaps the core of his own inspiration as a preacher, was to create the conditions in which more persons could attain such religious goals. Unfortunately, he never says this directly, and the substance of many of his writings is to the effect that in pursuing their religion the mass of ordinary Muslims, who will never become learned, are best advised to follow the example, teachings, and direction of the *ʿulamāʾ*.[22] He also seems to have seen initiation into a *ṭarīqa* as a privilege reserved for these same *ʿulamāʾ*.

Shaikh ʿUthman may have been proclaimed a *walī* early in his career by some of his followers, but his interest in his own *walāya* was closely related to the growing crisis which led to the *jihād*. Late in his life he wrote about two of his visionary experiences. The second of these, which occurred in 1794, was the more politically significant.[23] In it, several Muslim notables appeared to him, including the Prophet Muhammad and ʿAbd al-Qadir al-Jilani, founder of the Qādiriyya *ṭarīqa*. Al-Jilani was deputized by the others to bless the Shaikh's efforts to renew Islam and to gird him with the 'Sword of Truth'. As a result of this experience, ʿUthman saw himself as a legitimate *mujaddid*, although he was careful to dissociate himself from assertions by some that he was also the expected Mahdī. In his vision, the Shaikh was also authorized by al-Jilani to transmit a special *wird*, which he apparently passed on to many of his companions and followers in the *jihād*. The impact and influence of all these

[22] This view is expressed often in the Shaikh's writings. In *Tamyīz al-muslimīn min al-kāfirīn*, written some time after the *jihād*, he asserts that the only persons who are unequivocally Muslim are those who fall into one of three groups: the *ʿulamāʾ* whose belief is sound; the *ṭalaba* whose belief is sound; and those who listen to what the *ʿulamāʾ* say and who follow the example of the two above groups. See Uthman, *Bayān*, introduction, pp. 7–8.

[23] The Shaikh published an account of this vision late in his life in a work referred to as *Wird* (Zaria: Gaskiya Corporation, 1962). Relevant sections are translated in Hiskett, *The Sword of Truth*, p. 66.

aspects of Shaikh ʿUthman's *walāya* seem, however, to have been more relevant to the enhancement of the legitimacy of the *jihād* and of the new Caliphate than to the establishment of any *ṭarīqa* under his personal leadership.

## Ṭarīqa

Although Shaikh ʿUthman's name has become most closely associated with the Qādiriyya *ṭarīqa*, he was also affiliated with the Khalwatiyya and the Maḥmūdiyya (a branch of the Khalwatiyya founded in Air in the sixteenth century by Mahmud al-Baghdadi, and subsequently transmitted through Tuareg *mashā'ikh*).[24] Some authors suggest that he also acquired a Shādhilī affiliation;[25] and late in his life he received an initiation into the Qādiriyya-Mukhtāriyya.[26] Despite this wide range of *ṭarīqa* affiliations, the Shaikh was generally regarded by his contemporaries, as well as by subsequent historians, as a Qādiri Sufi, a perception which grew out of his close spiritual relationship with ʿAbd al-Qadir al-Jilani. Al-Jilani became a kind of 'patron saint' for the jihadist community, and the Qādiriyya became the 'established' *ṭarīqa* of the Caliphate. But this was not to the exclusion of other affiliations. As late as 1817, the year of Shaikh ʿUthman's death, his brother Abdullah wrote a treatise on the principles of Sufism, in which he emphasized the Khalwati initiation which he and ʿUthman had received from their *shaikh* Jibril b. ʿUmar.[27] No specific mention of the Qādiriyya appears in this text, although there is a long passage which elaborates the need to obey and remain loyal to one's *shaikh*. It seems unlikely that Abdullah intended this book as an implied attack on Shaikh ʿUthman. One must conclude that Abdullah did not view *ṭarīqa* affiliation as exclusivist, nor think that special devotion to al-Jilani should be emphasized to the exclusion of other *salāsil* and *asnād*.

[24] See ʿUthman b. Fudi, *al-Salāsil al-dhahabiyya li'l-sadāt al-ṣūfiyya*, University of Ibadan, MS 82/114, undated but probably written well before the *jihād*.

[25] Last, *The Sokoto Caliphate*, p. 6 n.; Martin, *Muslim Brotherhoods*, p. 18.

[26] This affiliation was noted in his *Taʿlīm al-ikhwān*, written in 1813; see B. G. Martin, 'Unbelief in the Western Sudan: ʿUthman dan Fodio's "Taʿlīm al-ikhwān"', *Middle East Studies*, 4 (1967), 93.

[27] ʿAbd Allah b. Fudi, *Bayān al-arkān wa'l-shurūṭ li'l-ṭarīqat al-ṣūfiyya, wa talqīn al-asmā' al-sabʿ ʿalā ṭarīqat al-sadāt al-Khalwatiyya*, University of Ibadan, MS 82/271.

Considerable evidence exists to suggest that Shaikh 'Uthman held a similar view of *ṭarīqa. Taṣawwuf* for him was a highly personalized religious science through which one sought the attainment of internal purification. The *ṭuruq*, with their supererogatory prayers and spiritual exercises, provided a specific method for pursuing this quest. Through the *salāsil* and *asnād* of the *ṭuruq* one established spiritual contact with the deceased *awliyā'* and with the Prophet himself, who thereby lent their personal blessings and support to one's efforts. One of the Shaikh's undated works clearly reflects this view of the *ṭuruq*.[28] In this brief treatise he not only gives the *salāsil* of his authorizations to recite the *shahāda* according to the Qādiriyya, the Khalwatiyya, and the Maḥmūdiyya, but he also includes a number of prayers without *salāsil* which could aid in one's personal devotions. The Shaikh collected such prayers and *salāsil*, and his behaviour reflects a widespread emphasis among African Muslims of that era upon the spiritual benefits and efficacy of recited prayers and mystical poetry. This process of collection was a highly personalized procedure; prayers derived their efficacy not only from the deceased saints who had composed them, but also from the religious stature of the persons who transmitted them.

A comparison of the attitudes of Shaikh 'Uthman and Sidi al-Mukhtar towards the collection and dissemination of prayers and *salāsil* demonstrates one of the fundamental differences which distinguish their relationships to the *ṭuruq*. They would have shared a sense of *taṣawwuf* as a personal spiritual quest; but Sidi al-Mukhtar, as *shaikh al-ṭarīqa al-Qādiriyya*, had established himself as the peerless Sufi leader of the southern Sahara. As such he no longer sought prayers and initiations; he dispensed them. Shaikh 'Uthman, by contrast, even as late as 1812–13, some years after the *jihād* had been successful in the Hausa states, accepted an initiation into the Qādiriyya-Mukhtāriyya from one of al-Mukhtar's *talāmīdh*.[29] This initiation strongly suggests that even if the Shaikh viewed himself as a dispenser of spiritual benefits, he was also still seeking them for himself.

The Shaikh's personal and devotional form of Sufism might

[28] *Al-Salāsil al-dhahabiyya.*

[29] This was Alfa Nuh b. al-Tahir al-Fullani al-Masini; see Martin, 'Unbelief', p. 93; Hunwick, *Sharī'a in Songhay*, p. 48.

have been transformed into a social and political organization based upon the concept of *ṭarīqa*, especially with the widespread recognition of his *walāya* and the successes of the *jihād*, but he made no determined efforts in this direction. He did not assert the uniqueness either of his own Sufi leadership or of any particular *ṭarīqa*, although he did consider himself to be a *mujaddid*. Rather, he adopted Abbasid models of state structure as the organizing principles of the new Caliphate, with himself as *amīr al-mu'minīn*.

It is probable that because of the Shaikh's close relationship to ʿAbd al-Qadir al-Jilani a sense of Qādiri identity pervaded the Caliphate leadership, and that this feeling was intensified after the death of the Shaikh through efforts to consecrate his memory by his son and successor, Muhammad Bello. But the crystallization of the Qādiriyya as an identifiable corporate group came only later, with the appearance of the Tijāniyya *ṭarīqa* in Sokoto, and especially after the death of Bello in 1837, with the controversy about whether or not he had accepted a Tijāni initiation from al-Hajj ʿUmar, who had spent some eight years in Sokoto.[30] That this possible affiliation aroused controversy at all was due specifically to Tijāni doctrinal exclusivity, which asserted the supremacy of this *ṭarīqa* over all others and demanded that its adherents reject all previous *ṭarīqa* affiliations. Bello's alleged initiation into the Tijāniyya therefore implied a rejection of both Shaikh ʿUthman and ʿAbd al-Qadir al-Jilani. The details of the debate which ensued in Sokoto do not concern us here; its very occurrence indicates how fundamentally Sufi concepts of *ṭarīqa* had changed in the region since the late eighteenth century. Whereas Shaikh ʿUthman had devotedly gathered *salāsil* from several *ṭuruq*, in the face of Tijāni exclusivity, the Qādiri leaders of Sokoto in the mid-nineteenth century began to develop their own corporate identity based upon the *walāya* of the Shaikh.

This transformation was accomplished by means of a concerted campaign led by Gidado dan Laima, *wazīr* to the now deceased Bello, and Nana Asma'u, daughter of Shaikh ʿUthman, wife of Gidado, and a highly trained scholar in her own right. In Arabic, Fulfulde, and Hausa, they wrote and

[30] See Last, *The Sokoto Caliphate*, pp. 216 ff.; J. N. Paden, *Religion and Political Culture in Kano* (Berkeley: University of California Press, 1973), 76–9.

disseminated treatises and poetry designed to sanctify the memory of the Shaikh as a saint.[31] They enumerated his *karāmāt* and endorsed the idea that his tomb was a sacred place of pilgrimage. Great emphasis was placed on the Qādiriyya as a *ṭarīqa* of unique qualities, and upon the spiritual attainments of Shaikh ʿUthman. One of the major long-term results of this campaign was the politicization of the Qādiriyya *ṭarīqa* in the Caliphate.

Evidence from their own writings suggests that Sidi al-Mukhtar and Shaikh ʿUthman shared similar concepts of *taṣawwuf* and *ṭarīqa* as the pursuit of a religious rule. They both sought to actualize the principles of Islam in their lives by entering into close relationships with Sufi *shaikhs* who transmitted to them a repertoire of litanies and prayers, and who directed them in spiritual exercises designed to support their efforts. Both also were accorded the status of *walī* by their contemporaries, due to their exceptional scholarly attainments, their exemplary religious behaviour, their visions, and their *karāmāt*. And they both employed the putative powers of their *walāya* as central elements in their efforts to reform society and polity. However, the *ṭarīqa* organizations which formed around each of them differed considerably from one another. These differences can be attributed to two principal factors: variations in their individual ideas about *ṭarīqa* as a corporate organization, and differences in the socio-economic milieux in which they operated.

Sidi al-Mukhtar placed great emphasis upon what he claimed to be the unique qualities of the Qādiriyya-Mukhtāriyya. Although he did not demand the exclusive loyalty of his acolytes in the way that the Tijānis would do, his assertions were not so different from theirs. Sidi al-Mukhtar insisted that he and his *ṭarīqa* constituted an unequalled source of spiritual benefit for his contemporaries. This attitude, which pervades much of his writings, is perhaps best symbolized by the fact that during his entire life Sidi al-Mukhtar acquired only a single Sufi initiation, from his teacher Sidi ʿAli b. Najib. This *silsila* linked him to ʿAbd al-Qadir al-Jilani and the Prophet, through the mediation

[31] See Last, *The Sokoto Caliphate*, pp. 210–13; J. Boyd and M. Last, 'The Role of Women as *Agents Religieux* in Sokoto', *Canadian Journal of African Studies*, 19/2 (1985), 283–300.

of various illustrious Kunta ancestors and Islamic notables, in particular Muhammad al-Maghili and Jalal al-Din al-Suyuti, two fifteenth-century scholars whose influence had been widely felt in West Africa and the Sahara.[32] By contrast, Shaikh ʿUthman received several different initiations into the Qādiriyya, including that into the Qādiriyya-Mukhtāriyya late in his life; and although he too was deeply devoted to ʿAbd al-Qadir al-Jilani, he also accepted affiliations with at least three other *ṭuruq*. Such differences suggest a wide disparity in how these two men viewed their role and status as Sufis. Sidi al-Mukhtar, much more than Shaikh ʿUthman, sought to centralize the *ṭarīqa* upon his person and to employ its organizational structure and ideological power in the reformation of the Saharan political economy. Shaikh ʿUthman, although he eventually published his visions and disseminated the *wird* which had been revealed to him, placed much less political emphasis on his Sufi attainment, which consequently retained more of a personal and private quality.

This relative absence of politicization in the Sufism of Shaikh ʿUthman also resulted from the fact that he and his associates in the *jihād* were preoccupied with building a state government based upon classical Islamic models. Sufi relationships within the *ṭarīqa* could aid in cementing loyalties within the new Caliphate, but predominating perceptions of *taṣawwuf* as a personal endeavour militated against the use of a Sufi model of *ṭarīqa* as a basis for political organization. The jihadists had decided to substitute an explicitly and patently legitimate Islamic state structure for the discredited Hausa governments which they had defeated; having made this political choice, and having argued so long and hard for the re-establishment of *sunna*, they were bound to follow the guide-lines laid down by the classical works to which they made reference. Great reliance was placed on these new offices of state in the effort to unify the disparate elements in the new Caliphate, and Sufism played no explicit role in the establishment of these new institutions.

Sidi al-Mukhtar created the Qādiriyya-Mukhtāriyya in a completely different context: most importantly, he had no

[32] This *silsila* is reproduced in Batran, Thesis, p. 420; on pp. 65 ff. Batran discusses the difficulties of verifying the historicity of the *silsila*. See also Hunwick, *Sharīʿa in Songhay*, pp. 42 ff.

intention of establishing a state structure and was therefore not bound by Islamic political precedents. On the contrary, attendance at his *zāwiya*, like membership of his *ṭarīqa*, was entirely voluntary, and considerable efforts were made to distinguish *ṭarīqa* activities from the enforced demands of state structures. Sidi al-Mukhtar established his organization in the interstices of the Saharan political economy, where he could find the greatest scope for innovation and growth. He employed the prestige of his *walāya* and of the Qādiriyya-Mukhtāriyya to cement Kunta loyalties, to establish and extend trade relationships, to amass individual and group clients, and to attract a widening network of Muslim scholars as his students. The *ṭarīqa* had the potential to transcend all existing structures and loyalties; indeed the Qādiriyya-Mukhtāriyya challenged many political, economic, and religious institutions in the desert, and in some cases took their place.

*Taṣawwuf* and *ṭariqa*, both as concepts and as institutions, contained enormous transformative potential, personal, social, and political. This potential was explicitly recognized in Sufi literature, and widely accepted by Muslims of all social categories in eighteenth- and nineteenth-century West Africa. As this discussion has demonstrated, the ways in which this transformative power might be employed, once it was acquired, varied among Sufi leaders depending upon numerous contextual factors. The *awliyā'* could affect social reality, and they in turn were affected by it. It was the interplay of such forces which led to the emergence of increasingly divergent forms of *ṭarīqa* organization in the nineteenth and twentieth centuries.

# 3

# *Khalwa* and the Career of Sainthood An Interpretative Essay

JEAN-LOUIS TRIAUD

THE term *charisma* is borrowed from the Apostle St Paul and refers to a grace, a free gift from God, bestowed upon an individual so that he may serve the community. An apparently close equivalent in Islamic tradition is the concept of *baraka*. The term means 'a benign force, of divine origin, which bestows physical superabundance and prosperity, and psychological happiness' (*Encyclopédie de l'Islam*). In the Qur'ān 'the term is never used in the singular, but only in the plural, and God is the unique and direct source of all the *barakāt*'.[1]

The effects of charisma and *baraka* are felt by those who come into contact with people who are endowed with them; but charisma, a providential gift, is attached to a person while *baraka*, a bestowal of force, can be transferred. It should also be noted that someone with Christian charisma would be entitled to a particular service, even a ministry, in the primitive community. However, charisma did not generally lead to sainthood, i.e. a state of perfection governed by proximity to the divine presence. *Baraka*, on the other hand, goes with sainthood; it helps in the working of miracles (*karāmāt*). Thus, the two concepts are not identical, despite analogies and common features. Nevertheless, the notion of *charismatic power*, as defined in this century by Max Weber in *Wirtschaft und Gesellschaft*, can be usefully applied to a number of holders of *baraka*.

Where does charisma come from? The element of personal gift (what Jean Séguy refers to as the 'seed') and the socially engendered element,[2] have rightly often been assessed. But there

We are grateful to our colleague and friend Sidi Mohamed Mahibou, Director for Arabic Education in the Republic of Niger, who helped us better to understand chapter 44 of the *Rimāḥ* of al-Hajj 'Umar, the text of which is often elliptical and difficult.

[1] H. A. R. Gibb, *Les Structures de la pensée religieuse de l'Islam* (Paris: Larose, 1950), 21.

[2] 'La Fabrique des Saints' is the title of Jean-Claude Schmitt's commentary on a collection of studies devoted to the cult of saints and to hagiographies in medieval Western Europe, in *Annales ESC* (Mar.–Apr. 1984), 286–300.

has been less awareness of the existence of an actual *career of sainthood*, subject to certain rules and following a well defined course. We have decided to discuss the *technology of charisma* in order to describe a number of codified practices which contribute to the sanctification of a postulant.

One practice is of particular interest: mystic retreat. At first sight, *khalwa* appears to be merely a very personal experience, a quest for the divine presence. But though its frequent occurrence in various Muslim regions, including West Africa, makes it more than simply a marginal phenomenon, scientific literature pays little attention to it. The odd remark acknowledges its existence; but *khalwa* is treated as an episodic matter of little consequence.

As a hypothesis, we should like to defend another point of view. It would be an excellent exercise—which we are, alas, unable to carry out—to record all occurrences of *khalwa*, both from written and oral evidence, in the context of West African Islam.[3] (This would probably reveal that *khalwa* is a much more

[3] Here are a few references from different regions:

On al-Hajj ʿUmar al-Futi's practice of *khalwa* itself, see B. G. Martin, *Muslim Brotherhoods in Nineteenth-Century Africa* (Cambridge: Cambridge University Press, 1976), 85 (after Abdallahi Ali, *Kitāb al-ajāʾib al-Qādriya*, MS, Dakar, Fonds Curtin, fo. 20[a]). Henri Gaden's notes on Mohammadou Aliou Tyam show how ʿUmar, at different stages of his *jihād*, awaits divine directives in a state of *khalwa* (M. A. Tyam, *La Vie d'El Hadj Omar, Qacida en Poular*, ed. and trans. by H. Gaden (Paris: Institut d'Ethnologie, 1935), 24, 53, 75, 130).

On the practice of *khalwa* in Bobo-Dioulasso at the beginning of the 19th c., see B. Traoré, 'Le Processus d'islamisation à Bobo-Dioulasso jusqu'à la fin du XIXe siècle: Approche historique et sociologique', MA Thesis (University of Ouagadougou, 1984). The author mentions Kassamba, one of the few Islamized groups at the beginning of the 19th c.: 'They were the most feared of all the Karamogow in the region as they practised *khalwa*' (p. 93). He shows the importance at the time of esoteric Islamic practices, particularly that of *khalwa*, among the Saghanogho of Bobo-Dioulasso (p. 148), and among the Diaby of Samatiguila in the Ivory Coast (p. 93). One of the key Islamic leaders and proselytizers of Bobo-Dioulasso, now venerated as a saint, Sakidi Sanou (born soon after 1850) was a well known adept at *khalwa*, which he had probably learnt from his master at Dia (pp. 93, 148).

On the practice of *khalwa* among the Koyam, a nomadic group with a high degree of religious specialization, settled in the south-east of present day Niger in the 18th c., see the *Documents scientifiques de la mission Tilho (1906–1909)*, ii (Paris: Imprimerie Nationale, 1911), 397 n. 1: 'It was their custom every year to undertake a sort of religious retreat during which they ate once a day and only left their hut to pray (all the huts being arranged around the *shaikh*'s hut). This retreat, which lasted forty days, was abolished during the years of famine.' We thank Maikorema Zakari, Assistant Lecturer at the University of Niamey (Niger), for having drawn our attention to this text.

widespread technique than appears at first sight, and would even enable us to draw up a map of its distribution. We should also discover that the technique is part of the function and career of the clerisy.) In West Africa, as Stewart's work on the dominant Qādiriyya-Kunta tradition has shown,[4] legal studies and mystic practices were combined during the great era (from the seventeenth to the nineteenth century) into a single mode of recruitment and career pattern. Mystic practices were a necessary part of the intellectual baggage of a religious aspirant, rather as a thesis and other academic exercises enhance a university career. Promotion to the highest ranks is rightly conditioned by tests taken, efforts made, and risks undergone, and also by the social recognition needed for acceptance by one's peers. In our view, *khalwa* is one such test: at the same time a rite of passage and a means to acceptance and promotion.

In the absence of the systematic survey which alone might confirm or refute the suggestions made above, we have decided to analyse the *khalwa* manual edited by a nineteenth-century West African Islamic master, al-Hajj ʿUmar al-Futi.[5] As a counterpoint we have added data on the same subject, collected during a field trip to Air in 1976.

The *khalwa* manual, after a brief introduction from al-Hajj ʿUmar, lists the twenty-six conditions (*shurūṭ*) for *khalwa*. The text is not original; in his conclusion ʿUmar says: 'With a few exceptions, this chapter is extracted from the *Waṣāya al-Qudsiyya*

[4] C. C. Stewart; 'Southern Saharan Scholarship and the *Bilad al-Sudan*', *Journal of African History*, 17/1 (1976), 73–93.

[5] Al-Hajj ʿUmar al-Futi, *Kitāb Rimāḥ ḥizb al-Raḥīm ʿalā nuḥūr ḥizb al- Rajīm* ('The book of the spears of the Merciful against the throat of the Evil'), chapter 44. The *Kitāb Rimāḥ* is a basic textbook for the West African Tijāniyya, written by ʿUmar in 1845 at Jegunko (in Futa Jalon). A printed edition was published in Cairo (1927, and reprinted) in the margins of the *Kitāb Jawāhir al-maʿānī wa bulūgh al-amānī*, a biography of Ahmad al-Tijāni by ʿAlī Ḥarāzim. We have relied on a similar edition published in Beirut (Dar al-Fikr, n.d.); for chapter 44, see pp. 153–60. An analysis of the *Kitāb Rimāḥ*, including passages devoted to *khalwa*, can be found in F. Dumont, *L'Anti-Sultan ou al-Hajj Omar Tal du Fouta, combattant de la foi* (Dakar and Abidjan: Les Nouvelles Éditions Africaines, 1974), 64–9. ʿUmar devotes two chapters to *khalwa*: 44 and 45. In this paper, we have only referred to ch. 44; ch. 45, which is shorter, discusses the personal preparation of the postulant, and distinguishes five kinds of *khalwa* (according to the length of time of the retreat and the invocations used, notably the introductory *sūra* (*al-Fātiḥa*), the *Basmala*, and the *al-Yāqūt al-Farīda* prayer).

and the *Khulāṣa al-Marḍiyya.*'[6] Thus, rather than innovate, ʿUmar remains within a long established tradition. A text of instructions does not, of course, reveal what happened in practice, so we shall not attempt to draw any definite conclusions from this analysis. Nevertheless, it provides what we might call the ideology of *khalwa* by defining its reasons, means, and aims in devotional thought.

From the beginning, the practice of *khalwa* is intimately linked to the master/disciple, *shaikh/ṭālib* relationship. ʿUmar's text often stresses this point, and there are several warnings against any attempt at 'savage' *khalwa*, without hierarchical and didactic control:[7] '*Khalwa* without a *shaikh* is a dangerous experience [*mukhāṭara*]' (introduction). 'The *khalwa* must be entered in the presence of the *shaikh* and under his mystic guidance [*amruhu al-bāṭin*]' (condition 2). 'The *shaikh* must enter the *khalwa* and achieve two *rakʿa* before the *murīd*'s retreat' (condition 5). The *murīd* must 'keep the *shaikh*'s image [*khayāl*] before his eyes, for the latter is his fellow-traveller [*rafīq*]' (condition 8). The *murīd* must 'continue to maintain the link [*ribāṭ*] which unites his heart [*qalb*] with the *shaikh*' (condition 20). Condition 21 is a reminder that the *shaikh* who has personal experience of *khalwa* ('he has experienced various states (*mārasa al-aḥwāl*) and overcome terrifying experiences (*rakiba al-aḥwāl*)') is the disciple's best instructor. He is like a doctor (*ṭabīb*) who knows what medicine will cure his patient. He is like the key which opens the door.

When the *murīd* sees unusual things, whether he is awake or asleep, he must also confide them to the *shaikh* 'without asking for an explanation [*ta'wīl*]; for the *shaikh* may not see any point in providing an explanation' (condition 23). The *murīd* must in no circumstances hide anything from him, since hiding (*kitmān*) something from one's *shaikh* is tantamount to treason (ibid.).

[6] *Risāla Waṣāya al-Qudsiyya:* advice to aspiring mystics, written in Jerusalem in 825/1422 by Abū Bakr al-Khawāfī Zain al-Dīn (757/1356–838/1435), who also lived in Cairo. See C. Brockelmann, *Geshichte der arabischen Literatur* (hereafter referred to as *GAL*), ii. 206. *Al-Khulāṣa al-Marḍiyya:* a book on mysticism, written by Shams al-Dīn Abū ʿAbdallah Muḥammad al-Malikī al-Madyanī (841/1437–881/1476) in Cairo. As the title indicates, this is a résumé (*khulāsa*) of an earlier work, *al-Durra al-mudīʾa*, about which we have no other information. See *GAL*, S. II, 151.

[7] The risk of madness is currently evoked in West Africa in warning against incautious indulgence in *khalwa*.

The *murīd* must neither change *shaikh*, nor divide his allegiance: 'if a *murīd* likes a *shaikh* who is not his own, his inner [*bāṭin*] spirit will not be able to reach the presence of God' (condition 20). ʿUmar goes on to explain that one cannot change *shaikh* any more than one can change the direction of the *Kaʿba*. All prophets are sent by God, but only the Prophet Muhammad can be asked to act as an intermediary; and each *shaikh* represents the Prophet in his own place. While all *shaikh*s are equal, one may address none but one's own *shaikh* (ibid.). 'Any opposition must be avoided, both to God and to the *shaikh*' (condition 21).

Thus the *shaikh* offers his disciples, and all believers, a human model inspired by analogy with and imitation of the charismatic personality *par excellence* in Islam, the Prophet Muhammad. The *shaikh*, and even more so the saint, somehow re-embodies the Prophetic model. As we shall see, the theories behind *khalwa* lay claim to the same model. Islamic charisma originates with Muhammad and speaks with his voice.

Mystic retreat has often been perceived as a flight from the world. This appears an incomplete definition: for on the contrary the retreat, given method and system, is part of a series of entirely worldly strategies. Only an élite, which can thereby be selected and recognized, can reach this level. Learning to cope with solitude and deprivation is less a flight from the world than an exercise in self-control—through solitude and lack of sleep and food. As ʿUmar writes in his introduction to the chapter, *khalwa* 'does not suit all people; everyone needs to experience the state [*ḥāl*] to which it brings him'. If *khalwa* was but a simple flight from the world, the text would not repeatedly warn against the lowest material appetites. It seems that in the eyes of many postulants *khalwa* is part of a strategy of power, which is why they are attracted to it. In condition 25 ʿUmar demands that one 'throw off—totally—all false appearance (*riyā*') and any striving for fame (*sumʿa*)'. He cites the case of a *murīd* who wanted to become a great orator when seeking *khalwa* but who became a plaything of Satan until his *shaikh* helped him back to the right road.

It is the taste for *karāmāt* which gives rise to the strictest warnings. The desire for *karāmāt* is in fact one of the most important motives for a number of the candidates who seek to enter *khalwa*. 'All the spiritual guides [*murshid*] have tried to spare the *murīd* the temptation of *karāmāt* by comparing this to a menstruation for men [*ḥaiḍ al-rijāl*]' (condition 7). The *murīd* must not, writes ʿUmar, enter *khalwa* with the aim of discovering [*kashf*] something about this world or material [*ʿainiyya*] *karāmāt*' (condition 25).

Here the historian's profane glance has a somewhat reductionist effect; and it might seem to the mystic engaged in a spiritual experience that this way of perceiving *khalwa* misses its essence. However, in the context of the period in question, we think that a social interpretation of the phenomenon is both useful and necessary. The triple test of isolation, lack of sleep, and lack of food acts as a means of selection. Only someone able to conquer his gross appetites and the urge to their immediate satisfaction, and who has given evidence of a divine presence (visions for example, but not to excess), is deemed worthy to join the 'club' of the *shaikh*s. This professional aim is never referred to as such in ʿUmar's handbook, though it constitutes one of the great underlying motivations in seeking the state of *khalwa*. In this context, one of our more important interlocutors during our enquiries in Air said: 'Someone who devotes forty days to God will have a pure heart and enlightened speech. *He will be listened to by everyone. Everything he says will be respected.*'[8] The candidate with the dignity of a *shaikh* must be distinguishable from the vulgar miracle-worker, from the seeker of easy profit. If he passes these different tests he acquires the rank of *shaikh*; and his peers, as well as ordinary believers, will recognize this. He may henceforth embark on a career of sainthood. *Karāmāt* and other blessings will then be accorded or attributed to him in addition.

But before that stage is reached, the test of *khalwa* must be completed. *Khalwa* is in fact an initiation test. The entire ritual (isolation, silence, deprivation of light, of food, and of sleep) leaves one in a state of near-death, a symbolic death. There is a constant tendency in mystic literature, whether Islamic or not, to compare retreat to a burial. The *murīd* must 'consider *khalwa*

[8] From interviews conducted at Egandawil (Air, Niger), 13 Apr. 1976, with the family of Malam Moussa, founder of the local *khalwatī* community. About this figure, see J. L. Triaud, 'Hommes de religion et confréries islamiques dans une société en crise, l'Air aux XIX^e et XX^e siècles: Le cas de la Khalwatiyya', *Cahiers d'études africaines*, 23(3)/91 (1988), 239–80.

as if it were his tomb [*ka-innahā qabruhu*]' (condition 4), and this 'until the Day of Resurrection' (condition 26). It is this voyage to the frontier between life and death which puts the *murīd* in touch with the sacred, from which he will derive authority and legitimacy if he is a man of power.

The candidate's training during the test concentrates on four aspects outlined in the first condition:

(1) lack of sleep (*sahar*);
(2) recital of the *dhikr*;
(3) restriction of food;
(4) isolation (*'uzla*).

In each of these practices the manual teaches and recommends moderation. *Khalwa* should involve a mastering of the senses and passions, not lead to uncontrollable excess. Extremism in these practices would show lack of maturity and hence imply failure.

### *Lack of Sleep* (sahar)

Condition 18 is entirely devoted to this question: 'The *murīd* must not sleep unless he has to, the limit being reached when he can no longer recite his *dhikr*'. If he manages to go without sleep, then 'his heart [*qalb*] loses the veil [*ḥijāb*],' which prevents him from seeing the fundamental truths; 'thus he sees the Divine Kingdom [*malakūt*] with the eyes of his heart'.

### *The Recital of the* dhikr

'Umar returns to this subject several times. The recital of the *dhikr* is the only activity of the *murīd* who totally devotes himself to it and meditates on its meaning during the entire test (conditions 6 and 9). Reciting the *dhikr* helps prevent the intrusion of personal thoughts. Generally, according to 'Umar, beginners are advised to ignore everything that comes into their minds, in order that they should not become a prey to their own thoughts (condition 19).

In condition 24, which once more invites the *murīd* 'constantly to pursue the *dhikr*', 'Umar evokes the authority of the founder of the Tijāniyya, Ahmad al-Tijāni, 'his *shaikh* and intermediary [*waṣīla*] with God', to distinguish between two kinds of *dhikr*:

'First there is the *dhikr* which frees the spirit [*rūḥ*] from all veils [*ḥijāb*]; then the *dhikr* which only partly succeeds, and which only removes one type of veil.'

The first category includes the following *adhkār* (plural of *dhikr*): *Lā ilāha illā -Llāh* ('There is no god but God'—the first part of the *shahāda*); *Subḥāna-Llāh* ('Glory to God'); *Al-Ḥamdu li-Llāh* ('Praise be to God'); *Basmala* (the act of saying '*bi-smi-Llāh*', 'In the name of God'); *Allāh, Allāh, Allāh*; *Allāh lā ilāha illā Huwa al-Ḥayy al-Qayyum* ('Allah, there is no god but him, the Living, the Everlasting'); *Al-Ṣalāt 'alā Rasūl Allāh* ('The Prayer to God's messenger'—'may God bless him and give him salvation!').

The second category includes 'the most beautiful names of God' (*al-asmā' al-ḥuṣnā*); 'each name removes a part of the veil, but not the other part'.[9]

### *The Restriction of Food: The Fast* (ṣaum)

The *murīd* must 'continually fast, which helps diminish the elements earth and water [*al-ajzā' al-ṭurabiyya wa-l-mā'iyya*]'—which make up the body—'and cleans the heart [*qalb*] of muddiness [*kadar*]' (condition 10). If the *murīd* has to eat, he may do it between the two evening prayers (*maghrib* and *'ishā'*); but it would be even better if he postponed eating as long as possible and waited for the moment of *saḥūr* (the last meal before sunrise) (ibid.). The state of *khalwa* is thus accompanied by the reinforced rules of the month of Ramadan, whatever period of the year it is.

This question of food also provides an example of that excess which should be avoided by a débutant impatient to reach the summit of the exercise: 'He must', writes 'Umar in condition 17, 'adopt moderation in the question of food: he must not have more than he needs, nor [allow himself to be subject to] excessive hunger.' Quoting al-Ghazali, 'Umar reiterates what is the hallmark of Sunnism, the happy medium (*wasaṭ*): 'for the best things belong to the happy medium'. Al-Ghazali even goes so far as to say that the demands expressed in the texts should not always be taken literally. So the *murīd* is not asked to mortify his body (and here the difference from ascetic practices is clear): he

[9] On the 99 'most beautiful names of God', see L. Gardet, *Encyclopédie de l'Islam*, 2nd edition, article *al-Asmā' al-ḥuṣnā*.

must 'forget his stomach' (*yansā baṭnahu*), the better to devote himself to the adoration of God. The ideal would be to become more like the angels (*malā'ika*), who feel neither the weight of the food on their stomach nor the pangs of hunger (ibid.).

### *Isolation* (ʿuzla)

The prime function of *khalwa* is complete separation from surrounding people. ʿUmar does not dwell long on the problem of the place of retreat. In his introduction he indicates that people mostly perform *khalwa* in a mosque; but it is well known that a hut, a cave, or any other isolated place is as suitable. In Air natural caves were generally used. According to collected evidence every village and every old mosque had a cave reserved for the purpose. But, as we were also told, 'If one wants to practise, one may create a special place.'[10] So where there is no cave one might use a building made of mud. 'The room for *khalwa* must be dark: neither sunrays nor daylight must enter' (condition 11). In fact, darkness is, apart from isolation, the most distinctive feature of *khalwa*. This point was frequently referred to by our interlocutors in the field: 'You need forty days in a dark place' or 'twenty, thirty, or forty days without light'.[11] The *murīd* 'should not lean on the wall during *khalwa*, nor should he rest his elbow on anything' (condition 8). He should bend his head and close his eyes while reciting the *dhikr* (ibid.), doubtless so as to attain sustained concentration and detachment. 'He must also continually follow the ritual ablutions [*wuḍū*]', which symbolically illuminate the *khalwa* (condition 12). He must allow no visitor access to the *khalwa* during his exercise (condition 22).

The *murīd* may leave the *khalwa* for two reasons only: the need to make his ritual ablutions and the need to pray (condition 15).[12] Indeed the *murīd* is advised not to miss the communal prayer, especially on Fridays (condition 16). It is difficult to reconcile the link with the community with the demand for strict

[10] Egandawil, interview (13 Apr. 1976).

[11] Ibid.

[12] He may also go out at night, to perform natural functions, as I was told at Egandawil.

solitude; quoting Suhrawardi,[13] ʿUmar suggests his preferred compromise solution: if it worries the *murīd* to go out, 'he should take someone along to pray with him during *khalwa*. He should not be content to pray alone; for one must fear for someone who abandons the communal prayer [*ṣalāt al-jamāʿa*]' (condition 16). When the *murīd* comes out, 'he should bow his head towards the ground without looking at anyone, and take special care to avoid the eyes of others' (condition 15). He should also cover his head and neck to avoid the 'sweat of the *dhikr* [*ʿaraq al-dhikr*]' (ibid.).

The solitude is naturally linked with silence. Condition 13 is entirely concerned with the question of silence (*sukūt*). Apart from the *dhikr*, discussions with the *shaikh*, prayer, and a few exceptional needs, the *murīd* 'must not utter a word': 'Each time he pronounces a useless word, he abandons the radiance [*nūrāniyya*] of his heart.' Equally, *khalwa* should take place 'away from all words uttered by men' (condition 14), which might distract the *murīd* and make him aware of the things of this world.

The ideal length of time for *khalwa* depends on the circumstances and varies from author to author. In his introduction, ʿUmar talks of periods of ten, thirty, and forty days—each of these numbers being derived from a prescription of the Prophet and the Qur'ān:

The shortest retreat lasts ten days, [the length of time during which the Prophet] practised *iʿtikāf*.[14] Ten days reinforce the mystic state [*ḥāl*] of the *kāmil*[15] and allow others to make progress [*tarqiyya*]. Certainly the

[13] Abū Hafs al-Suhrawardī (dec. 632/1234), author of some widely known treatises on Sufism (see Gardet, *Encyclopédie de l'Islam*, 1st edition, iv). Suhrawardi's *ʿAwārif al-maʿarif* (Beirut, 1966) is an important recognized Islamic source on the subject of *khalwa*.

[14] *Iʿtikāf* and *khalwa* are very close terms. They both refer to a retreat. The *iʿtikaf* is a pious retreat, described in the works on *fiqh*, but unlike *khalwa* it is not a mystic exercise aimed at seeking spiritual states. On *iʿtikaf*, see Gardet, *Encyclopédie de l'Islam*, 1st edition, ii; and 2nd edition, iv. According to tradition Muhammad spent the last ten days of Ramadan in *iʿtikāf*.

[15] *Cf. al-Insān al-kāmil* ('the perfect human being') in the terminology of the Sufi masters (ʿAbd al-Karim Jili, Ibn al-Arabi). The term denotes someone who has reached the highest levels of mystic knowledge (*maʿrifa*). Condition 26 refers to an experienced mystic who has passed the initial stage of *khalwa* and reached a superior level: 'This is the beginning of the spiritual [*maʿnawiyya*] *khalwa*. Judging from his external appearance, he is a human being like any other; but in his thought he is with the *maʿārif*.'

origin [of this practice] goes back to the *iʿtikāf* of the Prophet, and its aim is the cleansing [*taṭhīr*] of the heart [*qalb*] from the impurities [*adnās*] which result from [*human*] intercourse [*mulābasa*] . . .

In the same introduction ʿUmar recalls Suhrawardi's and Bukhari's descriptions of how the Prophet's Revelation followed a period of retreat: first, the Prophet had visions, then he chose to retreat into solitude (*khalā*'): 'He retreated alone into the cave of Ḥira' and he spent many nights in acts of devotion. Thereafter he returned home and took food . . . He continued in this way until the Truth came to him. He was then in the Cave of Hira', and the Angel appeared and said to him: "Recite [*iqra*']!"'

The thirty-day period, like the forty-day period, is attributed to Moses. It is written in the Qur'ān (Sura 7: 12): 'We have made a pact with Musa [*Moses*] for thirty nights. We then added another ten nights so that the meeting with his Lord lasted forty nights.'[16] ʿUmar makes the following comments about this text: 'The *sunna* indicates a period of forty days according to the pact made with Musa . . . but the aim is really to reach thirty days since this was the initial pact.' He adds: 'Muhammad stayed a [*lunar*] month in Hira' as it is stated in Muslim.'

The significance of these details concerning the length of the retreat is mostly symbolic; as we can see, they form part of a larger justificatory debate where the authority of the Qur'ān and the *sunna* is invoked in order to legitimize the orthodoxy of the practice itself. Such is the beginning of ʿUmar's instructions, which thus strictly follow the usual procedure of the *ʿulamā*'. Apart from these scriptural references which can be seen to have this other purpose, no absolute norms are given for the length of time. The postulants are even expressly advised not to fix a time limit in advance: 'The *murīd* should not fix a period after which he should come out [of *khalwa*] since this would make him think of the end of the period wishing it will come as soon as possible'

[16] The number forty has a highly symbolic value: 'And the flood was forty days upon the earth' (Gen. 7: 17); 'Then was Jesus led up of the Spirit into the wilderness to be tempted of the devil. And when he had fasted forty days and forty nights, he was afterward an hungered.' (Matt. 4: 1–2. See also Mark 1: 12–13; Luke 4: 1–2.) Hence in the Catholic Church Lent lasted for forty days. It may be noted also that from the 18th c. a 'quarantine' was imposed on travellers suspected of harbouring a contagious disease.

(condition 26). In practice forty or even thirty days are rarely achieved; practical difficulties impose a shorter period. The gradual nature of the training programme also encourages shorter but repeated periods of *khalwa*. According to Traoré's data, *khalwa* in Bobo-Dioulasso 'might last from seven to forty days. However, the *Karamogows*' retreat did not generally exceed a week . . . Indeed, the most common form is that of the *wouroudiri* (*wird*), which implies merely a night's retreat.'[17]

This then is ʿUmar's manual on *khalwa*, written in the middle of the nineteenth century. We have stressed the social function of the exercise. We believe that *khalwa* was of great importance to the West African clerisy between the seventeenth and the beginning of the twentieth centuries, in the greatest period of expansion of Sufi teachings and of the brotherhoods. For *khalwa* was a form of training which gave access to the higher echelons of the clerical profession and opened the door to a possible career of sainthood.

This splendid period is now over. The phenomenon has fared differently in different regions: in some it is quite simply extinct; in others it has been vulgarized. Only research in the field can throw light on a contemporary topography of *khalwa*. Our discussions in Egandawil revealed several times that *khalwa* was no longer practised and that no one used the places reserved for it any longer. In contrast, *khalwa* has become current *marabout* practice in Senegal. It is used in response to a demand: the *marabout* offers to perform it on behalf of the believers in return for payment. Thus *khalwa* is very close to the practice known as *istikhāra*.[18] Performing *khalwa* is a distinct feature of the clerical career. The *marabout*'s best disciples have access to it; it is part of the (usually oral) teaching provided by the master to his

[17] Traoré, 'Le Processus d'islamisation à Bobo-Dioulasso', p. 149.

[18] *Istikhāra* is a technique to solicit a divine response to a question of immediate interest. The person who carries it out generally lies down in either an ordinary or a holy place. The divine inspiration may be communicated in a dream. On *istikhāra* see Gardet, *Encyclopédie de l'Islam*, 2nd edition, iv. 271–2. As we have seen, ʿUmar himself periodically resorted to *khalwa* during *jihāds*; on the use of *istikhāra* and *khalwa* during military operations, see the notes in H. J. Fisher, 'Prayer and Military Activity in the History of Muslim Africa South of the Sahara', *Journal of African History*, 12/3 (1971), 393; 397–8.

disciple.[19] In view of such variation, any study on *khalwa* in West Africa should distinguish between different periods and different regions, with the aim of charting the phenomenon chronologically and geographically.[20]

We have deliberately said little about the mystical experience itself, as it is essentially incommunicable and impossible to describe. A historian's approach leaves study of the subject incomplete and must be supplemented by other means. What happens during *khalwa* touches the very depths of the individual and his personality; it is no mere response to the requirements of social appearances. In his introduction ʿUmar writes that *khalwa* leads to a profound (*ʿaẓīm*) discovery (*futūḥ*). We would note again that this dimension—which is not especially Islamic, for the technique of mystic retreat spans the common Judaeo-Islamo-Christian tradition—is, in our view, neither a pious deception nor a simple question of social convenience.[21] Nor is *khalwa* timeless and meaningless. We have tried to show how it is part of a set of social strategies. Essentially a clerical privilege, it enhances the social reputation of those who achieve it; despite the denials and condemnations contained in the manual, its main aim is often the social promotion of the postulant.

As is apparent among certain peoples at present, the vulgarization of *khalwa* limits its social effects. This contrasts with what we know of the seventeenth and eighteenth centuries, when its rarity enhanced its mystery—mystery being an essential ingredient of charisma. Though only historical case studies can confirm this hypothesis, *khalwa* appears to have been

[19] We thank Dr Mamadou Dia, a Senegalese scholar, for his useful remarks about current Senegalese practice.

[20] This has been done with other esoteric practices apparently common in West African Islam between the 17th and early 20th centuries; see L. Brenner, 'The Esoteric Sciences in West African Islam' in I. H. Abdalla and B. M. Du Toit (eds.), *African Healing Strategies* (Buffalo, NY: Trado-Medic Books, 1985). Note also D. M. Last's pioneering article: 'A Note on Attitudes to the Supernatural in the Sokoto *jihād*', *Journal of the Historical Society of Nigeria*, 4/1 (Dec. 1967), 3–13, which drew attention to the interest and importance of the supernatural in the study of Islamic societies in northern Nigeria.

[21] On *khalwa* and spiritual states and experiences, it is useful to study Laleh Bakhtiar, *Sufi Expressions of the Mystic Quest* (London: Thames and Hudson, 1976); also Michel Chodkiewicz, *Le Sceau des Saints. Prophetie et Saintete dans la Doctrine d'Ibn Arabi* (Paris: Gallimard, 1986).

part of an actual academic preparation for sainthood.[22] Having passed the test, the *murīd* took the rank of *shaikh*. He probably needed yet one more quality to become a saint (*walī*). But that is another story . . .

[22] Just as we were about to finish this text, we came across a recent article which provides an additional argument in support of our point of view: Hassan Elboudrari, 'Quand les saints font les villes: Lecture anthropologique de la pratique sociale d'un saint marocain du XVII^e siècle', *Annales ESC* (May–June 1985). Discussing the itinerary of the saint Mawlay ʿAbdallah al-Sharif (1596–1678), the author recounts how the life of this figure was described in the hagiographic tradition: 'The account of the saint's success as handed down in the hagiographic tradition makes this appear as *the logical culmination of a continued ascent*. In this text the achievement closes a *cycle* which has proceeded smoothly through the different stages. Thus, the young candidate to sainthood passes through a series of peregrinations . . . before settling in a small hamlet (of the country). There he embarks on an ascetic life of intense meditation away from the world, in retreat (*khalwa*) from his surroundings. This is where he finally achieves 'spiritual openness' (*fath.*) and receives the 'injunction' from the Prophet, which *confers sainthood upon him. His fame is instant* and he is unanimously recognized' (p. 491; our italics). As the author goes on to say, the hagiographies invariably gloss over any point of doubt or contradiction in such an edifying life, while creating, according to a recognized procedure, an official formalized biography. However, what is interesting here is precisely the fact that this official biography is built around a model, a curriculum, or a career in which *khalwa* plays a crucial part. Hence this example, from another Islamic region, conforms to the pattern which we have attempted to outline in this paper.

For this chapter we gratefully acknowledge the translation of Mrs Mette McRae.

# 4

# Charisma and the Crisis of Power in East Africa

FRANÇOIS CONSTANTIN

> With the exception of Judaism and Protestantism, all religions and religious ethics have had to reintroduce cults of saints, heroes and functional gods in order to accommodate themselves to the needs of the masses.
>
> MAX WEBER

> *Man lā Shaikha lahu fa 'l-Shaiṭānu shaykhuhu*
>
> AL MAYMANI

> Ce n'est pas ce qu'on fait qui compte, c'est l'histoire, c'est l'histoire qu'on raconte.
>
> YVES DUTHEIL

THOUGH charisma is a personal gift, it exists in the minds of other people. One does not preach in the desert. Thus, though we are discussing an essentially universal social phenomenon, its success depends on the special nature of the society where it appears. Prophets and saints are not exclusive to West Africa or the Maghreb, even if very elaborate forms of Sufism have been developed there. Though more modest, certainly less well known, they exist on the shores of the Indian Ocean too, and even in the heart of Central Africa.[1]

In fact the East African context is paradoxical. Islam arrived there earlier than elsewhere in sub-Saharan Africa (the first century after the *hijra*); but it long remained strictly confined to that small social élite of Arab origins who were the dominant class on the east coast of Africa. So our focus is upon an Islamic frontier where, by the end of the nineteenth century, new *awliyā'*

[1] There are, of course, some 'great' saints on the East African coast. They have been studied by B. G. Martin in 'Muslim Politics and Resistance to Colonial Rule: Shaikh Uways b. Muhammad al-Barāwī and the Qādiriyya Brotherhood in East Africa', *Journal of African History*, 10/3 (1969), 471–86; and 'The Qādiriyya and Shādhiliyya Brotherhoods in East Africa, 1880–1910', in *Muslim Brotherhoods in Nineteenth-Century Africa* (Cambridge: Cambridge University Press, 1976), 152–76.

appeared to shocked glances from the 'organic' *'ulamā'*[2] of the Omani or Hadrami ruling groups. Here, more than elsewhere perhaps, the brotherhoods led by indigenous *shaikhs*[3] became a popular religious force and a haven for the disinherited, who, reduced to social subordination by official state dogma, rediscovered the dignity they needed for social redemption.[4]

Dogmatic or racist theories must be avoided. Islamic charisma is not a form of blind antichristian or anti-European fanaticism. It became the motor for a kind of reaction against what was perceived as oppression even when exercised by a Muslim authority.[5] Thus the Qādiriyya, the Shādhiliyya,[6] and others such as the Husseiniyya, contributed to an Islamization of the indigenous peoples and slaves never contemplated by the ruling Arabs, whose monopoly over the Arab–Islamic culture was a cornerstone of their position of quasi-hegemony.[7] Accosted and seduced by *shaikhs* of the brotherhoods, the Africans, whether slaves or not, could acquire certain crucial attributes of the ruling class (religion, symbolism, ritual, and language—even culture), a first step towards formulating a

[2] This description is inspired by Gramsci and refers to a category of intellectuals who have put their learning at the service of the ruling class.

[3] The brotherhood message had been introduced by *shaikhs* from Somalia (e.g. Shaikh Uways) or the Comores (Shaikh Maaruf).

[4] See C. Coulon, *Les Musulmans et le pouvoir en Afrique noire* (Paris: Karthala, 1983); also F. Cooper, 'Islam and Cultural Hegemony: The Ideology of Slave Owners on the East African Coast', in P. Lovejoy (ed.), *The Ideology of Slavery in Africa* (London: Sage Publications, 1981), 271–307.

[5] This at least is what is suggested in studies on activism in East African brotherhoods; see A. P. Caplan, *Choice and Constraint in a Swahili Community* (Oxford: Oxford University Press, 1975); R. C. Greenstein, 'Shaykhs and *ṭarīqa*s: The Early Muslim *'ulamā'* and *ṭarīqa* Development in Malawi, 1885–1949', History Research Seminar paper (University of Malawi, 1976/7); A. Hjort, *Savanna Town: Rural Ties and Urban Opportunities in Northern Kenya* (Stockholm: Stockholm Studies in Social Anthropology, 1979); A. H. Nimtz, 'The Role of the Muslim Sufi Order in Political Change', Ph. D. thesis (Indiana University, 1973); id., *Islam and Politics in East Africa: The Sufi Order in Tanzania* (Minneapolis: University of Minnesota Press, 1980). On the fringe of the region under study there is also U. Braukamper, 'Notes on the Islamicization and the Muslim Shrines on the Harar Plateau', in T. Labahn (ed.), *Proceedings of the Second International Congress of Somali Studies* (Hamburg: H. Buske Verlag, 1984), ii. 145–74.

[6] The Qādiriyya and the Shādhiliyya arrived on the coast around 1880. There is an older and quite different brotherhood tradition in the north (Somalia and Ethiopia).

[7] This desire to adhere to Islam can be interpreted as a sign of the success of the ruling class, showing how subordinated people seek integration into the culture of the ruler ('He loved Big Brother'); cf. M. M. Mulokozi, 'Revolution and Reaction in Swahili Poetry', *Utafiti*, 1/2 (1976), 127–47.

demand for the right to equality and self-determination. But in a world where nothing was clear-cut such subversion took place quietly and with due respect for the established order, even under colonial rule.[8] It was a calm charismatic force, which became significant because of its wide appeal.

A sense of nuance, or moderation of purpose and conduct, rarely characterizes the political or religious leader who can mobilize the masses. So how has this strange recent phenomenon gained ground among the brotherhoods? Flexible, decentralized, and politically moderate, the brotherhoods have evolved among social groups where the means of production of charisma (the ideological apparatus)[9] appears to be out of reach of the *shaikh*s and their disciples. This weakness in the control of charismatic technology may have contributed considerably to the appreciation of their religious and social eminence precisely because the Arab rulers thought that they did not need to strengthen their ideological supremacy. In this context, where media manipulation is non-existent, it may be possible to discover the essence of charisma, the 'seed' without which this idiosyncratic phenomenon would be doomed to failure. We may also gain a clearer assessment than elsewhere of the significance of the context in which it flourishes—*the ecology of charisma*—and thus reach a better understanding of its social importance.

## 1. IN SEARCH OF THE SEED OF SAINTHOOD

In the sociological analysis of ideologies, rational attempts to explain the irrational provide much material for thought. But this is often circumscribed by two attitudes both of which lack scientific rigour. According to one God exists and miracles are in the nature of things, since a chosen few are invested with divine powers. The only problem is how to distinguish the select few

[8] The brotherhoods were only partly involved in the disturbances which marked the anti-European struggle, whether religious (the letters from Mecca) or political. (Rumaliza was not a zealous religious chief but a slave trader whose profits had declined. The intrigues at the Bugandan court developed before the brotherhood *shaikh*s had landed on the coast).

[9] That is to say traditional forms of propaganda: *griot* or writing, hagiography, and other ideological tools which apparently had a weak impact here.

from the charlatans.[10] The other, claiming to be rationalistic and materialistic, relegates dreams, visions, and miracles to that bag of tricks which rulers employ to take advantage of the masses, aggravating their alienation by giving them heroes and saints to worship. God or no God, alienation or no alienation, the fact remains that faith exists and functions and that the inexplicable may have considerable social consequences.[11] It is the reason for irrationality which must be elucidated; for it is present in all societies from the most archaic to the most sophisticated, from the medicine man to General de Gaulle.

### *A Few Notes on Personalized Power*

To believe, said the philosopher Paul Ricoeur, is to 'take to be true'. Faith is a banal, common social phenomenon which operates in the sphere of imagery and fantasy and often reflects an inability to grasp reality. The risk of 'false' consciousness is considerable. Be that as it may, such imagery is itself a reality with which we must come to terms. A sociologist or a student of politics cannot ignore what every demagogue and politician knows. The miracles of saints or visionaries are possibly or probably successful feats of manipulation; but we still need to understand how they succeed.

This is where charisma comes in. It exists; and everyone has encountered it. In all social groups there are individuals who attract attention and to whom people listen. It is this individual influence that social psychology is trying to analyse. It is said of some that they have authority (because of their wisdom, their respect for the rules, and their efficiency), of some that they dominate (through control over economic resources or means of production, or simply through physical strength or access to arms), and of others that they have special powers (they accomplish, or appear to accomplish, extraordinary acts,

[10] In institutionalized religions this is a justification for canonization processes (which are not exempt from profane manipulation). However, although judged by theological criteria (whatever their validity) a 'charlatan' does not deserve sainthood, he still deserves sociological attention as soon as his manipulations and tricks have a mobilizing effect. This is the case with the cranks, mystics, or wily demagogues studied by N. Cohn in *The Pursuit of the Millennium* (London: Secker and Warburg, 1957).

[11] This is one of the limitations of mechanistic Marxism, which Gramsci's 'discovery' has allowed us to overcome.

beyond the reach of ordinary mortals).[12] We think of this last category when we talk of charisma, which is seen as a special gift; those confronted with the supernatural seek explanation in some divine presence. Identifying the signs of charisma is much easier when the person in question practises mysticism, takes things to excess, or generally behaves in such a way as to widen the gap which separates him from ordinary mortals. In a very rigid society merely being different from the norm may be enough.

The East African *shaikhs* who are venerated today are of this last type. They are remarkable personalities (if colonial administrators' reports, the collective memory, and contemporary researchers are to be believed), who initially acquired social standing by adopting the daily practices required of all good Muslims. This was enough to make them appear nonconformist renegades, likely to foment disorder, to the dignitaries of the Islamic institutions of the Zanzibar Sultanate (who practised an exclusive form of Islam, rejecting all proselytism to ensure—so they imagined—the perpetuation of their power).[13] This latently seditious devoutness presupposed a strong personality and moral and intellectual qualities out of the ordinary, or at any rate exceptional enough to arouse the attention of the masses and inspire a charismatic relationship. But this is still too vague to provide the basis for the durable reputation necessary for veneration and sainthood. The 'seed' of charisma is made up not only of factors related to both individual and relational religious qualities, but also (for the two spheres cannot be separated in Islam) of social factors,[14] which then give the *shaikh*'s word total authority.[15]

[12] See J.-W. Lapierre, *Le Pouvoir politique* (Paris: Presses Universitaires de France, 1969).

[13] This may explain why J. S. Trimingham appears to have become blind to humble East African *shaikhs* after observing influential *marabouts* and *khalīfas* in West Africa; see his *Islam in East Africa* (Oxford: Clarendon Press, 1964).

[14] It is time to refer back to Weber, specifically to his *Sociology of Religion* (Boston: Beacon Press, 1964); and also to J. Séguy, 'Charisme, sacerdoce, fondation: Autour de L. M. Grignion de Montfort', *Social Compass*, 29/1 (1982), 5–24; id., 'Pour une sociologie de l'ordre religieux', *Archives des sciences sociales des religions*, 57/1 (Jan.–Mar. 1984), 55–68.

[15] The term is used in the sociological sense developed by P. Bourdieu; see his *Ce que parler veut dire* (Paris: Fayard, 1982).

### *Religious Resources*

The major *shaikhs* who spread the message of the Qādiriyya or the Shādhiliyya and became *walīs* did not engage in any mystic-military struggle or *jihād* against the colonial power (which they knew as well as, if not better than, the West African *marabouts*).[16] They were neither Mahdīs nor 'fanatics', still less Horsemen of the Apocalypse. They launched neither devastating revolts nor military *jihāds*. They were pious men, determined to spread the Prophet's word to those who did not yet know it, i.e. the vast majority of the people of the Zanj. By personal choice as much as by necessity they felt that their missionary duties did not oblige them to conduct wars, whether holy or secular, even though there was much infidelity or heterodoxy to eradicate. Their only weapon was the word (of the Qur'ān); of the various models or types of charismatic leader, they could have been seen as *wise men*.[17] This image of wisdom was perhaps inspired by the fact that the religious agents in question were of the kind called 'small' saints. They were not founders of new brotherhoods but members of a school which had been brought to East Africa by other saints, usually Arabs, from whom they had received *ijāza*.[18] However, their role was crucial; for it was they who spread Islam to the interior of the continent.

The brotherhood network provided a pillar of wisdom which complemented or, if need be, replaced other supporting assets such as scholarship, gifts, or birth. It also introduced the importance of the transnational in the production of charisma. For this network, or web of family links, involved the *shaikhs* in a kind of transnational charismatic organization by providing a permanent link with the cradle of Islam, the Arab Peninsula—the 'other shore' towards which so many East African eyes were turned. This link guaranteed the legitimacy and accuracy of the religious discourse which took place at Bagamoyo, Ujiji, Nkhota-Kota, or Lamu. Whatever the social background of the

[16] It should be recalled that in the east there was a double colonial domination, consisting of the Zanzibar Sultanate and the important Arab traders on the one hand and the European colonizers (with their administrators, soldiers, and bibles) on the other.

[17] By way of illustration, we should categorize them with Thierno Bokar rather than Shaikh Hamallah; they are certainly very far from Uthman b. Fudi.

[18] Nor were the 'great' saints (see n. 1) armed prophets.

*shaikh* (who might be of Maniema slave or of Hadrami descent), the fact that he was a link in a chain which connected him through time and space with the Prophet gave his words authority and made him an example for others. The pupil of the *shaikh*, his disciple, knew that he belonged to a large universal family, the *umma*, even though the local *'ulamā'* contested his claim to such integration.[19] Thus the disciple abandoned his servile, hard working, 'uncivilized' position (*mshenzi*) and expected to enter the world of the ruling class (*ustaarabu*).

Such promotion could occur outside the brotherhood as long as the master himself came from a well connected Islamic family. In Lamu Habib Saleh's pupils could not pretend to be related to their *mwalimu* since he was a *sharīf*, a direct descendant of the Prophet through Fatima, from a background of social and religious distinction which commanded great respect in East Africa. It gave Habib Saleh certain inborn qualities (*baraka*, Islamic knowledge) which could only be transmitted through direct descent. Moreover, he belonged to the Ba'Alawi clan from whose ranks is chosen the *quṭb al-aqṭāb*, the spiritual chief of all the Hadramis and the only person who can intercede directly with Allah. Though Habib Saleh could not include his disciples in his family group[20] the 'transnational family organization' to which he also belonged helped to legitimize his religious initiatives and the innovations which he introduced in the poor parts of Lamu, and to stifle the criticisms of the local *waungwana* leaders.[21]

Although no one was as cosmopolitan as Habib Saleh (born in the Comores), all the venerated *shaikh*s we know travelled.

[19] The Zanzibar *'ulamā'* had only a small audience consisting of important Arabs; these were the only people in whom they were interested. As far as the Omanis were concerned, they belonged to the Khariji branch, which had a very limited following, whereas brotherhoods such as the Qādiriyya had a cosmopolitan, even universal, dimension.

[20] Habib Saleh's sharifian lineage was a closed group. This was why it was considered an order rather than a brotherhood. Cf. A. H. M. El Zein, *The Sacred Meadows: A Structural Analysis of Religious Symbolism in an East African Town, Lamu* (Evanston: Northwestern University Press, 1974).

[21] The innovations introduced in the Lamu ritual were inspired by practices from the Hadramawt (this was the basis of their legitimacy, inhibiting criticism from the important Hadramis at Lamu), which is also where Habib Saleh found the name for the mosque he constructed. See El Zein, *The Sacred Meadows*; A. W. Boyd, 'To Praise the Prophet: A Processual Symbolic Analysis of "*Maulidi*", a Muslim Ritual in Lamu, Kenya', Ph.D. thesis (Indiana University, 1981).

Their mobility seems to have added to their charisma. They rarely went to Mecca, but they all visited what was locally considered the end of the accessible world: Zanzibar, widely known as a centre for Islamic scholarship. All are known at some stage to have studied at one of the island's *madrasas*, supervised by *shaikhs* of international fame. A *shaikh* who left his family and home to study gave evidence of the sincerity of his religious vocation, thus enhancing his prestige.[22]

Apart from allowing the *shaikhs* to claim membership of a multinational network, these journeys offered proof of their Islamic knowledge (*ʿilm*). Such knowledge is evidently relative, and the criteria for being thought learned were not in 1920 the same on the shores of Lake Malawi as in Zanzibar or Medina. The *shaikhs* of Bagamoyo, Lindi, or Nkhota-Kota were not considered authoritative theologians or *ʿulamāʾ* throughout the intellectual world of Islam; but they knew enough to challenge the local élites, and sometimes acquired recognized administrative authority.[23] The schools they opened had pupils from far away, occasionally from overseas. Their religious competence was judged by the *ijāza* they had received.[24] The *shaikh*'s reputation as a scholar depended upon the name of his teacher; the Zanzibar *madrasas* were particularly well thought of. The transfer of scholarship within the brotherhood enhanced the prestige of the neophyte, since his knowledge was acquired through a chain of transmission (*silsila*) which went back to the Prophet and thus short-circuited the local hierarchies. Moreover, the vagaries of colonial policy sometimes inadvertently confirmed the *shaikhs*' competence in Muslim sciences. Fearing trouble, the German administration in 1909 systematically searched the houses of the main *walimu* of Tanganyika, where

[22] Mystic exercises are often associated with religious virtuosity and used as personal tests; but the *shaikhs* we are studying do not appear to have attempted the kind of mystic-masochist exercise which has made West African or Ethiopian saints famous. *Khalwa* is unknown; it is certainly not practised.

[23] For instance the colonial administration conferred on Shaikh Ramiya, the former Maniema slave, the title of *Liwali* in Bagamoyo. For his part Habib Saleh enhanced his fame by using his learning to cast ridicule on the so-called local intellectuals.

[24] Habib Saleh, for instance, had double legitimacy, having received not only the *ijāza* of a Mungwana *shaikh* from Lamu but also that of the *quṭb al-aqṭāb*, the spiritual leader of all the Hadramis; see El Zein, *The Sacred Meadows*, esp. p. 138. In the same way, Shaikh Ramiya's initial success was based on the fact that he had received the *ijāza* of a famous Arab *shaikh*, Shaikh Hussayn al-Lughani.

they found an impressive selection of books from all corners of the Muslim world, covering all areas of Islamic sciences.[25] So the *shaikh*s were not illiterates, contrary to what was thought in Europe and Zanzibar. Such bureaucratic actions only confirmed their local reputation.

*ʿIlm* was impressive, but hardly enough to make one appear out of the ordinary. It might be complemented by sharifian birth; but to be considered exceptional one needed to prove that in addition to these human qualities one had been given the blessing of God: the ability to work miracles (*karāmāt*) and the possession of *baraka* were signs of 'pure' charisma. The proofs are there, in popular accounts and official reports. The edifying life-stories of *shaikh*s of humble origins like Shaikh Ramiya or Shaikh Binti Mtumwa already appeared near-miraculous;[26] even so, social achievement did not automatically confer religious prestige. It is not clear whether the public recognition of actual miracles preceded or succeeded the acknowledgement of social success;[27] but the miracles happened. Shaikh Ramiya made his disciples walk on water. Habib Saleh could stop storms or blazing fires. Shaikh Kiponda b. Selemani, on the banks of the Malawi, predicted future events (such as the rains, the First World War, and the Spanish flu). What is interesting about the well-known miracles is their social dimension: the *shaikh*'s *baraka*, like his knowledge, was employed for the good of his followers.[28]

It was not merely the collective memory of these miraculous feats that provided evidence of the *shaikh*s' altruism. It could also be observed in their exercise of more routine functions which were still the preserve of men who had special links with God

[25] This was how C. Becker was able to draw up his comprehensive bibliography; see his 'Materials for the Understanding of Islam in German East Africa', *Tanzania Notes and Records*, 68 (1968), 31–61.

[26] Once more birth was important, but by way of contrast. Incidentally, when Shaikh Ramiya died the lawyers in charge of determining the succession found it hard to admit that an old *mshenzi* could have left a more substantial legacy than many Arabs; they thereby unwittingly provided a bureaucratic confirmation of the miraculous nature of the Shaikh's life.

[27] The public may have heard of miracles performed in youth at a later date; the timelag between the event and its becoming public knowledge must of course be taken into consideration.

[28] This same *baraka* may have served to sow panic among the *shaikh*'s adversaries by definition bad Muslims.

and the spirits: I am referring to therapeutic or psychotherapeutic techniques. Some *shaikh*s, such as Habib Saleh, were famous for their competence in Islamic medicine. Others, as A. P. Caplan notes,[29] intervened in possession cults like shamans, helping the patients (often women) with charms, talismans, or special rituals based on Muslim 'material'. However, nothing is known of the healing powers of those who became *walī*s.[30]

It seems that the exceptional gift, that special condition which alone could inspire subsequent veneration, was linked to social conduct which defied the norms of the conventional pecking order.

## *Social Popularity*

If Weber is right in suggesting that the cult of saints responds to the needs of the masses, it is easier to understand why those who witnessed the feats of the venerated *shaikh*s paid such attention to their social conduct and their treatment of others. In a highly stratified society,[31] the masses tend, more or less consciously, to strive for more humane treatment and a reduction of barriers between the rulers and the ruled. They long for a modicum of dignity and respect and want the religious practice of the dominant groups to be based upon the supposedly generous and universal principles of Islam, whether this be under Omani or Hadrami rulers, with Gujerati or Baloush agents, with Swahili perhaps or Shirazi followers claiming Arab origins. Whether of humble origins (*mshenzi*) or a *sharīf*, the future *walī* provided a link between the various social categories which he himself transcended, from political leaders to slaves.

His daily life was lived according to the rules of Islam—which did not preclude material wealth; on the contrary, this was an extra sign of grace. His private life was blameless, so far as it was publicly scrutinized (in a small community, certain things are hard to hide). As student, husband, and father he was admirable; and he fulfilled his religious duties with as much zeal

[29] Caplan, *Choice and Constraint in a Swahili Community*, pp. 118–23.

[30] For the ordinary shaman, this is strictly a kind of charisma of office. None of the saints whom we have studied seems to have had a special reputation as a healer.

[31] See part 2 below.

as he devoted to his social ones. Thus Shaikh Ramiya continued to hold his erstwhile master—admired for his generosity—in great respect even after he reached a high position in the administration. In contrast, the ascension to sainthood of Shaikh Haji Mkwenda was obstructed by his amorous affairs. And Shaikh Himedi wa Phanzi's shady political manœuvres against Shaikh Binti Mtumwa, whose private life was exemplary, actually precipitated the decline of the Qādiriyya brotherhood in the Lake Malawi region.

The road to sainthood demands not only an exemplary private life but also generosity to others. The *shaikh* must help his followers to become good Muslims and prepare themselves for the day of divine judgement. Hence he must set up schools to teach the Qur'ān to anyone who does not know it, regardless of age or social and ethnic origins. He should pay particular attention to the underprivileged, generally forgotten because of their povery (*washenzi*) and unable to pay school fees. Though it is only to be expected from a pious man, such selfless action enhances the *shaikh*'s reputation as an exceptional person.[32] Anyone who has all the attributes needed to belong to a social élite, but who renounces such distinction, is a non-conformist whose attitude is incomprehensible to the establishment but widely admired by subordinate groups, and therefore by the majority.[33]

Those were the people the future *walī* taught to read the Book and the words of the Prophet. Although he used texts by Arab mystics he made sure the message was communicated, whereas the discourse of the *'ulamā'* had the sole function of consolidating Omani power. Yet such a *shaikh* did not become a political agitator. He did not seek short term influence. Secular power was not for him. He accepted the established authority and considered it his duty not only to intercede with God on behalf of those who trusted him but also, in the shorter term, to intercede with the Prince. The latter could not easily repress him, both because he was a *shaikh* and because he showed no inclination to

[32] Note Habib Saleh's reply to the local dignitaries who came to ask him to put his learning at the service of his own class: it is those who live in ignorance who must first hear the Word of God.

[33] We are again referring to the multiple pressures (seduction, corruption, and threats) which the Hadrami dignitaries brought to bear on Habib Saleh to make him resume his rank; nevertheless he did not make a noble (or socially desirable) marriage.

challenge the colonial authority. He merely tried to minimize injustice in this world and to prepare for life in the next.[34]

This is not to deny that many saints amassed great wealth in this world. Some, particularly Hadrami or Comoran *shaikhs*, inherited family fortunes. But those from humbler backgrounds started with very little (perhaps a hand-out from their master)—although they knew how to make their assets grow, usually in trade, transport, or plantations. The modest unostentatious lifestyle of a *walī* permitted the accumulation of resources which could be used both to finance pilgrimages or studies for his disciples and to help them materially. The system of clienteles centred on the *shaikh* demanded that he had the means to guarantee his followers' protection, both spiritual (by the running of the *madrasa*) and material (by gifts and financial support for certain family ceremonies).

This pragmatism, not to say business sense, had no adverse effects on the *shaikh*'s charisma, which in its ideal type was disinterested. In East Africa at least, destitution was not a precondition for acknowledged sainthood. In contrast to the ordinary shaman, the *shaikh* was indeed disinterested as far as his religious activities went. He did not profit from his gifts; he sold neither his *baraka* nor his religious protection. But the services expected from him necessitated an assured income, and they were financed prosaically—through the astute management of a business (in anything from coconuts to real estate), presents from local dignitaries seeking atonement, remuneration for certain administrative functions, or even pensions (as in the case of Shaikh Binti Mtumwa, a war widow). Such resources, added to gifts from the disciples, made the Qādiriyya branch at Bagamoyo 'a lucrative business venture'.[35] Evident entrepreneurial flair was perhaps, quite openly, a crucial element in the make-up of a charismatic person, as long as the dividends were reinvested to serve the right religious cause.[36] There evidently

[34] 'The former *shaikh* did everything he could to maintain peace among all parties, and his suggestions to the government corresponded to the people's wants.' A popular assessment of Shaikh Ramiya, as reported by Nimtz in 'The Role of the Muslim Sufi Order in Political Change', p. 455.

[35] Nimtz, *Islam and Politics in East Africa*, p. 178.

[36] And invested prudently; for it was a necessary result that the *shaikh*'s spending capacity increased.

existed a political economy of charisma, although we still know too little about it.

However, in order to consider this phenomenon we must go beyond individual abilities and the 'autopsies' of saints and examine the general context in which the seed of charisma can take root.

## 2. THE ECOLOGY OF CHARISMA: THE SOCIAL PRODUCTION OF SAINTHOOD

Pareto may have been wrong to suggest that history is a graveyard of aristocracies. But it is certainly a mass grave of geniuses whose personal qualities have failed to get historic recognition and who thus remain unknown. This is why we need to explain a phenomenon which at first sight appears fortuitous: why, despite the number of potential candidates, did so few East Africans exercise charismatic power during the period between 1880 and 1930? The production of saints cannot be circumscribed in time and space by sheer accident. Hence we must explore the environment in which these *shaikh*s achieved sainthood.[37]

To be admired one must have admirers. In other words, circumstances must favour a particular sensitivity towards the exceptional.[38] At certain periods certain social groups may be particularly exposed. At times of depression or uncertainty the need for dreams, magic, and immediate or eternal reassurance encourages their susceptibilities. Two sets of social factors seem to allow charisma to be recognized in a society: the first are of a structural nature, linked to collective life and derived from social psychology; the second, more circumstantial but no less important, are linked to the struggles for power which went on throughout East Africa in the nineteenth and twentieth centuries.

[37] This factor of environment is all the more important within Islam, where there is no bureaucratized procedure for sanctification, which results rather from a lengthy popular process which does not exclude manipulations.

[38] This is dependent upon the mechanics of the functioning of beliefs.

### *The Mass Psychology of Charisma*

Increased research into the mechanisms of power has led to a recognition of the significance of subjective phenomena which shape individual and collective attitudes and behaviour. Human nature's effective and emotive susceptibilities are now recognized as important variables in sociological analysis. However, what remains problematic is the question of how to explain these factors with scientific rigour. Without going as far as Reich,[39] sociologists of religion throw some light on the ideological functioning of mysticism and prophecy and on the sects who practise these.[40] Their observations, often made in a European context, are not very different from studies of charismatic Muslims; it seems that all religious practice involving élites and theologians shares certain elements. Since every society has its miracles, mysteries, and mystics it is tempting to suggest that collective mental structures exist which favour the germination of charisma. Depending on the relevant circumstances (such as culture) these latent structures come to play a more or less active part in shaping society.

Pierre Ansart, who has contributed to the rehabilitation of the subjective as politically decisive, has studied the way in which 'passion' intervenes and functions in social life. The 'communication between the individual and society'[41] takes place in that emotional sphere where message, persuasion, and propaganda are exchanged and ideology is transmitted. According to Ansart '[many] dimensions of political life happen at the level of collective sensibility'. Susceptibility to emotions is not a peculiar feature of 'traditional' societies or of subordinate classes, as the outmoded developmentalists claim. It is present everywhere, even if it manifests itself in different ways. 'Emotional identification is particularly strong among small groups . . . influenced by slightly different mechanisms than society as a whole . . .'.

A group of disciples around a *shaikh* is a case in point: here the

[39] We are referring to the analyses developed by W. Reich in *The Mass Psychology of Fascism* (New York: Farrar, Strauss and Giroux Inc., 1970).

[40] See especially Séguy, 'Pour une sociologie de l'ondre religieux'.

[41] See P. Ansart, *La Gestion des passions politiques* (Lausanne: L'Age d'Homme, 1983), 93–108. All quotations in this section are from this source.

lowly, the marginalized, and the oppressed find a way out by changing their miserable condition, at least psychologically. The *zāwiya* functions more or less clearly as an ideological group '[that] aims to denounce the established order . . . and pronounce the true law'. It is not a 'utopian group' like the sects which cut themselves off, turning in upon themselves to escape reality and create their own ideal world.[42] Opposition to the established order makes the ideological groups of the former type very demanding: their members' conduct must be blameless, guided by high moral notions. To internalize the new norms, they resort to conspicuous 'active demonstrations'. Apart from their ritual religious significance, the *maulid* and the *dhikr* act as a psycho-social stimulus. The *dhikr* illustrates the 'verbal exchanges within the group [which] continually repeats the evidence' in order to achieve adhesion and cohesion.[43]

The attempt to achieve freedom does not necessarily eliminate dependency. But this dependency is displaced and partly transferred to the group. The ideal is often incarnated and expressed by a leader and his image: 'the group depends upon its chief for its intellectual nourishment, and the chief expresses and integrates individual desires'; this is true of the *shaikh*.

The word, whether repetitive and communal or discursive and unilateral (the *shaikh*'s instruction), is more than merely a means of communication. It is the key to authority. In Islam, the Word is that of God himself. In this context the mechanisms for linguistic domination are particularly important in a society with an oral tradition. Pierre Bourdieu has shown the social effects of linguistic inequality.[44] Inequality in the use of language and in ability of expression ('he speaks well') is accentuated in a multilingual situation where one language is considered 'noble' and the others 'vulgar'. This was how, in East Africa, Arabic was seen in relation to other languages including

[42] Not every brotherhood practice should be regarded as an attempt at psychic withdrawal or flight from society (escapism).

[43] On these subjects Ansart develops the analysis of W. R. Bion, *Recherche sur les petits groupes* (Paris: Presses Universitaires de France, 1972).

[44] See especially *Questions de sociologie* (Paris: Éditions de Minuit, 1980) 96–125, from which all quotations in this passage are taken; also *Ce que parler veut dire* (cited above, n. 15).

Kiswahili.[45] Certain norms apply in the linguistic market of speakers and listeners: for example, the contrasts written/oral, sophisticated/vulgar, and intellectual/uneducated are used to identify authorized speakers (with the 'right' to speak and to command) and the modalities of authoritative speech. These norms, produced and internalized by society, maintain the power-relationship as long as the belief in the right to speak and the duty to listen continues to exist. Such relationships occur in the religious sphere too.[46] They presuppose the intervention of a legitimate speaker, 'someone who knows the laws of the system and who is therefore recognized and co-opted' (the *shaikh* with his *ijāza*) and legitimate listeners (the speaker selects his audience—the disciples), with a legitimate relationship (the disciples acknowledge the *shaikh*) and a legitimate language. Arabic is the language of the Qur'ān; but apart from the reading of the word of God the proceedings between *shaikh* and disciples must be intelligible and so must take place in a vulgar language (unthinkable for the *ʿulamā'*).

The *shaikh* from a humble background who had mastered Arabic (the language-tool of political, religious, and economic domination) had risen above the ordinary and was therefore heard. The linguistic product he controlled was valuable: it offered salvation in the next world, as well as immediate social rehabilitation and Muslim dignity—even for someone with poor Arabic.[47] Esoteric speech, which was essential but incomprehensible, had almost magical power.

Magic is a kind of manipulation which exists in Arab–Islamic cultures as well as in African (and other) cultures.[48] It is central

[45] The symbolic value of Kiswahili has always been important. Though a sign of the richness of African culture, it is played down by the 'organic' linguists of the dominant groups, who will only recognize Arabic or the European languages. Such tactics are to be expected in symbolic struggles.

[46] At the extreme one hears without either listening or understanding: 'The liturgy is a ritualized and totally codified language (gestures and words) which follows an entirely predictable order. The Latin liturgy is the extreme form of a language which is authorized but not understood, and thus nevertheless can function as a language to the satisfaction of both speakers and listeners . . . Its main effect [is] to make believe, to make respect, to make admit—to seek acceptance, even if one does not understand.' (Bourdieu, *Questions de Sociologie*, p. 101).

[47] In the same vein see B. Étienne, 'La Moelle de la prédication: Essai sur le prône politique dans l'Islam contemporain', *Revue francaise de science politique*, 33/4 (Aug. 1983), 706–20.

[48] And doubtless in all religions.

to practices used to heal individuals whose psychological disorders might upset the social order. Such therapy relies on possession cults. A. P. Caplan, studying social dynamics in the island of Mafia (Tanzania), encountered Qādiri associations which served both as guilds and as therapeutic centres for psychological disorders. The *mkobero* was run by a *mwalim* who acted as shaman, using tools and talismans of partially Islamic derivation in the presence of a congregation who recited the *dhikr* and chanted words supposedly in Arabic.[49] In this island the *mkobero* was only used by the lower orders, while another cult was run by shamans from the economic élite. The identity of such a congregation might explain the link between the brotherhoods and the possession ritual. It was among the Africans, especially the deprived, the lonely, and the women,[50] that the *shaikh* found his audience. His therapeutic powers—with mysterious techniques but obvious aims—were evidence of his charisma, resolving personality crises which jeopardized public order exactly because they betrayed its unacceptable aspects. So it is that features of Bantu society can be said at a certain stage to have facilitated the spread of Muslim charisma.

To summarize: the masses' psychological means of expression are in fact a function of their actual living conditions.

### *Social Crisis and the Activation of Charisma*

Every society has political passion, pays respect to the authorized Word, and recognizes the healing powers of psychotherapy. Sainthood does not simply arise as a result of the accumulation of a certain number of signs which can be listed and added up. Such an accumulation will only be activated under certain decisive civil and political circumstances. It can be asserted that the brotherhoods emerged at a time of deep structural crisis in East African society; hence we are faced with the classic issue of the providential man in relation to popular social movements.

[49] See Caplan, *Choice and Constraint in a Swahili Community*, pp. 118–19.

[50] See F. Constantin, 'Condition féminine et dynamique confrérique en Afrique orientale', *Islam et sociétés au sud du Sahara*, 1 (1987), 58–69.

(i) *The crisis of hegemony in East Africa*[51]

For a long time, power relations were more or less stable along the East African coast. Although for various reasons there were vicissitudes, the basic power structure remained unchanged for about a thousand years.

The small coastal principalities or city-states prospered, thanks to their control over commercial activities. They provided the existing international market (that of the Indian Ocean and the China Sea) with products from the interior (Zanj), namely gold, ivory, and slaves, which were sold on the coast. The rulers of these principalities, sultans of Arab (or supposedly Arab) origin controlled the business world without attempting to institutionalize their political power. These political and economic rulers had collaborators who, though of local origin, gradually absorbed certain elements of the Arab Muslim culture. They acted as business managers and actively supported the rulers' cult of their own superiority (*ustaarabu*, meaning 'civilization'). The rest of the society consisted of Africans who had come from the interior to seek work (or for other reasons) or who had been brought against their will to become slaves. They were considered to be uncivilized, savages (*washenzi*).

Coastal rivalries and political instability led to foreign intervention. First on the scene were neighbours who already had local interests, to wit the Omanis and Hadramis. They reconciled local disputants by imposing their political, economic, and military rule upon them all. Other foreign powers were called in by individual rulers in trouble. This gave the British and the Prussians a pretext for establishing themselves, and they, like their predecessors, squashed local rivalries by

[51] The bibliography related to this chapter of history is so rich that it is hard to provide a few references. Let us note the three massive volumes of the *History of East Africa* (Oxford: Clarendon Press, (i, ed. R. Oliver and G. Mathew) 1963, (ii, ed. V. Marlow and E. M. Chilver) 1965, (iii, ed. D. A. Low and A. Smith) 1976); and, at the other end of the scale, E. S. Atieno-Odhiambo, T. I. Ousu, and J. F. M. Williams; *A History of East Africa* (London: Longman, 1977). We have published a bibliography in F. Constantin, *Les Communautés musulmanes d'Afrique orientale* (Pau: CREPAO, 1983). For the coastal region, apart from works mentioned above, one may consult R. Pouwels, 'The Medieval Foundations of East African Islam', *International Journal of African Historical Studies*, 11/2 (1978), 201–26, and 11/3 (1978), 393–409; id., 'Islam and Islamic Leadership in the Coastal Communities of Eastern Africa, 1700–1914', Ph.D. thesis (University of California, 1979); id., *Horn and Crescent*.

introducing colonial rule. But there was no change in the status of the Africans, who remained serfs even when they were eventually liberated.[52] They stayed at the bottom of the social scale, with few rights and no privileges.

The political situation changed in about 1840. The city-states passed under the nominal authority of the Sultan of Oman, who transferred his capital to Zanzibar. Caravans were dispatched further and further into the interior to keep up with the demand for slaves on the international market. But such investment was only beginning to pay off when the Europeans arrived, with persuasive arguments which called a halt to the boom. The abolition, first of the slave trade, then of slavery itself, meant that the rulers needed a new source of profit. They transferred their investments to plantations; but the market was already dominated by Europeans. A thousand years of political and economic domination were at an end.

Even socially, the rulers had lost their sway; for their cultural hegemony had not been complete. For a thousand years, the Arab–Muslim way of life had been the ideal for the subordinated Africans—hence the evolution of the Swahili culture, also reserved for a minority. The religion of the Qur'ān had carried great prestige, not least because it was so exclusive. The rulers were aware that their cultural norms would be debased by being spread to the *washenzi*, and that their own superiority would thereby be eroded. It was not until the middle of the last century that Islamic caravans began to leave the narrow coastal strip and penetrate the interior, not to spread Islam, but in search of profit; though once installed they might influence their immediate neighbours and encourage a few conversions (these however remained on the individual level).[53] When Qādiriyya and Shādhiliyya *shaikh*s arrived on the coast later in the century, they became the first to undertake a serious programme of proselytism, which was no longer limited to local élites of a

[52] More than half a dozen treaties, spread over more than half a century, were needed before slavery was abolished by law. Resistance to abolition came from both 'producers' (mainly but not exclusively Arabs) and 'consumers', especially French traders and planters.

[53] It should be recalled that the Yao (in Malawi and northern Mozambique) were Islamized in a different way. A mass conversion, initiated by the Yao chiefs, showed radical opposition to colonial penetration of the interior.

broadly Arab or Swahili character. The élites were concerned about these missionary activities, over which they had no control, and which allowed the masses to appropriate the norms upon which they had based their own legitimacy. It may seem paradoxical that this process of Islamization, far from strengthening Arab hegemony, was actually eroding its base. In fact Arab power declined further because at the same time the Europeans arrived with their own Book, their missionaries, and their politico-military arguments.[54] If one had to choose a monotheistic religion, it was perhaps safest to invest in the one proposed by the imperial power.

There was plenty of scope for unrest in a society which had lived so long with the mirage of *ustaarabu* and which, just as this seemed within reach, found itself at odds with those it tried to imitate. Moreover, it was becoming clear that this splendid civilization lacked the (political) strength to stand up to the one which was now challenging it. The ideological basis for authority was evidently shattered; Islam (which had managed to keep its rifts hidden during the period of Arab rule) now revealed its heterogeneity. The Omani rulers' desire to monopolize the Kharijiyya did not prevent others from seeking instruction in different versions of the religion. Sunnism may not have been the Islam of the local élite, but it was a form of Islam—and reputedly the most widespread. The arrival of Asians, and increased overseas exchanges, introduced a further element of relativism into attitudes towards the dogmatism of the total legal and spiritual power exercised by the ruling *'ulamā'*.

A combination of factors in late nineteenth-century East Africa had thus created conditions particularly propitious for the emergence of Messianic or millenarian figures, or of dissident, independent, or Ethiopian churches.[55] In a stratified, racialist society (although this was not openly admitted, social status was inextricably linked to skin colour), religious equality is unthinkable. Discrimination (in the access to the priesthood

[54] It is true that the Portuguese had started to arrive from the end of the 15th c.; but on the whole they tended not to stay long in any one place.

[55] On this type of movement in Europe see Cohn, *The Pursuit of the Millennium*. A broader analysis is presented in V. Lanternari, *Les Mouvements religieux des peuples opprimés* (Paris: La Découverte Maspéro, 1983); English edition, trans. L. Sergio: The Religions of the Oppressed (New York: Alfred Knopf, 1963).

or other religious functions) leads neophytes to fall back upon their own resources. Failing to be admitted on an equal footing to the official religion, they create a special corner for themselves within it.[56] Islam with its lack of a centralized structure is especially open to such movements.

Although to a superficial examination order seemed to reign, civil society was in fact at crisis point. Each group that had arrived on the coast had imposed its authority and created new racial strata. And the indigenous people, the Africans, had remained at the bottom of the social ladder.[57] In the circumstances the appearance of spiritual guides was a matter of need rather than accident.

(ii) *Charisma and the new order*

A collapse of the social structure is difficult to cope with. When social order is shattered, when values are eroded or destroyed, and the individual feels himself to be without protection, material, legal, or moral, phenomena which enhance a personal sense of identity acquire a special significance. A protector, a prophet, or a *walī* may provide the means to recreate a semblance of structure, and thus of stability and security. Marc Bloch illustrated this process in his brilliant analysis of the consequences for Western Europe's political and social structures of the fall of Rome.[58] In feudal medieval times, the result was the triumph of personalized power and a surge of prophetism.[59] A system of patron–client relations often alleviates the effects of total social disorder.

The power-structure in East Africa had collapsed at the beginning of the twentieth century; but the *shaikh* provided continuity and coherence. He was the fixed point of a religious

[56] A broad outline of this process is offered by B. Hargrove (who quotes J. C. Hoekendijk on the situation in South Africa) in *The Sociology of Religion* (Arlington: AHM Publishing Corporation, 1979). But he makes the error of assuming that this kind of movement constitutes a return to a 'primitive' form of religion.

[57] The initial interactive Arab–African dualism was succeeded by a many-runged ladder with the Europeans on top, above the Asians; below them came the Arabs and finally the Africans. This hierarchy, which was established by law, was further complicated by the fact that there were different ranks within each category, with irregulars like the Waswahili on the margins.

[58] See M. Bloch, *La Société féodale* (Paris: Albin Michel, 1968); English edition, trans. L. A. Manyon: *Feudal Society* (London: Routledge and Kegan Paul, 1962).

[59] The historical process is analysed by Cohn in *The Pursuit of the Millennium*.

network which remained unaffected by the vicissitudes of local political life. For the Africans adherence to the *umma* meant social recognition and promotion.[60] Through his own person and the network he represented the *shaikh* recorded everyday life, created a social structure, and provided protection—in immediate material terms (the natural and social requirements for survival), in terms of order (the brotherhood provided shelter), and against dangers beyond the control of ordinary mortals, whether related to the struggle for temporal power of the 'great' or to rather more mysterious problems. In the last case the solution might involve magic or communication with the supernatural, in other words with God. An Arabic-speaking *shaikh* could talk to God.

Although the desire for order clearly has a strong affect on individual behaviour, there is a limit to the price people will pay for this. A return to total subordination is an intolerable price. The personal strength of the *shaikh* provided another way out. He was African, or at least not compromised by links with groups too closely related to the arbitrary, oppressive foreign administration. He constituted a meeting-point, a centre of mobilization open to anyone longing to replace the unequal established order. This reinforced the image of 'opposition movements' often associated with the brotherhoods.[61] But the concept should not be too rigidly or mechanically applied. A pious man does not at once enter into secular conflicts. He merely offers his disciples the means to correct their past errors and embark on the right Way (*ṭarīqa*); and he reinforces their psychological resistance to interference and aggression from an authority which they do not recognize. But he is not a mystic utopian who encourages them to retreat from the world, nor a naïve revolutionary who thinks his faithful fanatics can take on the ruling power. Though there have been anticolonial revolts since 1880, the brotherhoods have not really been involved in these.[62]

[60] The importance of the religious reference in the *shaikh*'s manner and speech doubtless explains why he resorts neither to demagogy nor to 'populism' (to the extent that this implies a vulgar message or lax style). The *shaikh* needed to maintain a social and symbolic distance from his disciples. The incautious use of the term 'populist', for example by B. G. Martin and R. Pouwels, may lead to confusion.

[61] For a convincing analysis in terms of *contre pouvoir* see Coulon, *Les Musulmans et le pouvoir*.

[62] See n. 8 above.

The gifted and scholarly *shaikh*, a religious virtuoso, tended to feel that his abilities should be better employed. He was aware that his status and fame forced the authorities to treat him carefully; one does not suppress a man of God with a large following, and administrators, whether colonial or not, preferred to avoid ructions. And he knew that the district officer distrusted him, watched him, and wrote reports on him for his records. But a sort of deal could be, and often was, concluded whereby the colonial power tolerated the *shaikh*, legitimized by his disciples and by his function. He could publicly express the concerns of those who, despite their lack of access to official institutions, constituted the majority: the Africans. In this way he served the authorities, who could take the appropriate measures to avert any risk of trouble and who could count on the *shaikh* to control his followers. Traditionally the administration considered the future saint a well-behaved radical.[63] He might even, like Shaikh Ramiya, join the administration, though without renouncing his role as spokesman, which he would now fulfil from within. In this way majority rule might be brought a tiny step nearer.[64]

Both in the short and in the long term the *shaikh* played the role of go-between, multi-purpose mediator, and broker between different parts of society. He helped break down the rigid barriers of society in order to advance its most deprived members in the eyes of both God and the Governor. He was thus neither revolutionary nor conservative. He spoke on behalf of the people, voicing their concerns and contributing to social stability by advocating adjustments or even reforms. The

[63] This overall assessment does not preclude the possibility of cruder opposition on the local level, as Habib Saleh or Shaikh Ramiya discovered. But the *shaikh* does not have the limited horizon of the local dignitaries. A crucial element in his prestige consists in being able to pass over their heads and address someone 'higher up', or even The Most High.

[64] Thus when Shaikh Ramiya died his son Muhammad became head of the Qādiriyya (a sign that the basis of authority was sliding), and when the Tanganyika African Association was set up Muhammad became its second member in Bagamoyo, an action which could only encourage his disciples to rally to the nationalist movement. It should be noted, however, that the founder of the local branch of the TAA was from the same background as Shaikh Ramiya, being a Maniema, a fact which suggests that the brotherhood movement's contribution to nationalism was more closely related to its popular African recruitment than to its religious characteristics; see Nimtz, *Islam and Politics in East Africa*.

importance attributed to this role enhanced the image of wisdom which the *shaikh*s were acquiring.

Having been the saint of the oppressed, the brotherhood *shaikh* appeared to lose his importance after independence. The development of nationalism and political parties seemed to indicate that political institutionalization signalled the end of his historical mission. If this is true, it follows that religious charisma had occurred mainly because of the significance of the racial question in socio-political conditions. But there are signs that the brotherhoods still exist and the shrines continue to be visited. There are still situations of dependence or alienation which can be analysed in terms similar to those employed in the past. The state power is often excessively rigid and obsessed with national unity. If its legitimacy were ever at risk, it might be tempted to incorporate the *umma* into a state-controlled, bureaucratic organization. Were this to happen, a new generation of *walīs* might well emerge, although they would be unlikely to be as well-behaved as those of the '*belle époque*'.

# 5

# Muslim Charisma in Burkina Faso

RENÉ OTAYEK

It is not easy to discuss Upper Volta, now Burkina Faso,[1] in a collective work about Islamic charisma, since the phenomenon appears at first sight to be locally nonexistent. Marginal study may be intellectually justified from a scientific point of view; but it may prove rather frustrating. Hence it is tempting to distort reality so as to adapt it at all costs to a prefabricated analytical pattern. However pertinent this pattern may be—and the notion of Muslim charisma *is* pertinent when the process of Islamization in sub-Saharan Africa is being analysed—it is inappropriate here, or at any rate incapable of reflecting the singularity and complexity of a situation where Islam has such peculiar features.

From this point of view, researchers in Burkinabe Islam must be permanently on guard against two types of attitude which, although contradictory, stem from the same distortion of reality. They both relate to the conditions of Islam's expansion in Burkina, which is a fairly recent phenomenon. These conditions, with which we shall deal, have given rise to a form of Islam which could not be assimilated to any other model, though certain of its characteristics can be found elsewhere south of the Sahara (in southern Ivory Coast and some parts of East Africa).

In view of our initial observation it may be tempting to consider Burkinabe Islam as a kind of Islam of a 'secondary zone'. Indeed colonial Islamic literature has long considered Burkina, especially Mossi, a 'bastion' against Islam. The progress of Burkina's Islamization has been systematically underestimated or quite simply ignored. This misguided attitude is largely due to the ideological presuppositions of the colonial administration, and has led to a series of works which

[1] For the sake of convenience the name Burkina Faso will be used throughout this text, though for all events before 4 August 1984 'the Republic of Upper Volta' would strictly have been correct.

have long continued, sometimes unconsciously, to promote the image of an entirely animist, or even Christian, Burkina. This is the first attitude to guard against.

The second attitude flows naturally alongside the first, despite the fact that it is its complete negation. It is just as misguided to consider Burkinabe Islam to have participated in the African 'Islamic revolution', whether silently or not. Islam there has not had the political influence it has had elsewhere. It does not have the intellectual vigour which has made neighbouring Islamic communities famous, nor the brotherhoods which give the Senegalese Mourides their notorious influence. Although expanding, Islam in Burkina is still seeking its rightful place within the *umma*. To accord it a doctrinal influence and a political dimension which it has not (so far) achieved would be misleading.

In other words a researcher should approach this form of Islam with an open mind, and without a priori assumptions derived from observations of neighbouring countries where Islam has developed differently.

One feature of Islam in Burkina—and an important one—is precisely that it has not produced truly charismatic figures, certainly not in the same proportion as elsewhere. This irrefutable fact is supported by all the available literature; the study of the Islamization process in the cultural space which became Burkina Faso in 1984 reveals little evidence of the exceptional, the supernatural, or the extraordinary. Miracles are rare. Burkina is not a favoured land of charisma.

However, scarcity does not equal total absence. In Burkina too Islam has occasionally produced personalities considered by their religious community to be out of the ordinary, people admired for their wisdom, piety, and scholarship, people venerated for their *baraka*. The crucial difference is that whereas elsewhere charisma was the rule it remained the exception in Burkina. It is as though Islam there continued to be on the margins of the 'market' of charisma—as though charisma was less in demand than elsewhere. And when, very occasionally, charismatic Muslims emerged in Burkina, they rarely attained the spiritual completeness, influence, or authority of Amadu Bamba, ʿUthman b. Fudi, al-Hajj ʿUmar, or Shaikh Hamallah.

Further observations are relevant to the consideration of charisma. First it should be stressed that the concept has no equivalent in Islam. The nearest is that of *baraka*. But the two concepts are not always identical. A man can have *baraka* without attracting that special popularity which shows that his extraordinary gifts are recognized. Moreover, *baraka* can be attached to a line of descent and thus be inherited, whereas charisma is a personal gift. Similarly it is difficult to define the concept of a saint within the sphere of Islamic religion. The Qur'ān ignores the notion of sainthood. And though *walī* is its nearest equivalent it is not an exact translation.

To these epistemological reservations, appended to those specifically concerning Burkina, we would add some cautionary questions. The ideal type of the charismatic person, as elaborated by Max Weber, is of course a tool of theoretical analysis. In fact the phenomenon is rarely found in its 'pure' form. The researcher's task is further complicated by the diversity of social types derived from it. How should one distinguish charisma, in a society with an oral tradition, from the simple respect commanded by literacy? In Islam true charisma is linked to the sacred Word as revealed by God in the Qur'ān. How can one distinguish charisma from the simple prestige conferred by Africans on the representative of a culture perceived as superior by virtue of the intrinsic mastery of reading and writing? Is the relationship between the *ṭālib* and the *mallam*, however erudite the latter, different in kind from a respectful student–teacher relationship? Are we not in danger of seeing charisma everywhere, even in cases of mere 'ordinary' relationships based upon a principle of authority which might be anything but charismatic?

These few remarks were intended to clarify the subsequent discussion by outlining the difficulties of identifying the phenomenon of charisma in sub-Saharan African societies. In the case of Burkina these difficulties are further compounded by a host of factors, historical, sociological, and political, because of which the charismatic phenomenon reveals itself hesitantly there.

## 1. HIDDEN CHARISMA

Much the most important of the factors which held back the growth of charisma in Burkina was the political structure of what was to become the independent Republic of Upper Volta. Centred around a group of societies which were particularly impervious to external influence (the Mossi kingdoms) it was a country where Islam had achieved only minor influence.

### *A Socially Marginal and Politically Weak Form of Islam*

It would be as misguided to deny the historical significance of the Mogho Naaba's empire for the political structure of modern Burkina as to ignore the importance of other ethnic and cultural factors. According to legend the Mossi empire was founded in the eleventh century AD, but total Mossi domination over the eastern and central regions of what is now Burkina was not established until the fifteenth century. Theories about the origins of the Mossi people vary; but is is generally agreed that they came from Gambaga, now in Ghana, and that they settled in an area where they rapidly assimilated the original inhabitants.[2] These had been organized in family or village groups until the Mossi warriors introduced an entirely new system of political organization: the State.[3]

The Mossi empire consisted of nineteen kingdoms. The most important were Tenkodogo, Yatanga, and of course Ouagadougou (Vogdogo), the Mogho Naaba's capital, which had nominal sovereignty over all the other kingdoms.[4] Political authority, conceived as divine, was concentrated in the hands of the Mogho Naaba, whose powers were limited only by considerations of respect for traditional customs and religious rules. The sovereign was the living symbol of a sacred chain which went back to the mythical founding fathers. The members of the royal family, under the delegation of the

[2] On the origins of the Mossi see, inter alia, M. Izard, *Introduction à l'histoire des royaumes Mossi*, 2 vols. (Paris and Ouagadougou: CNRS and CVRS, 1970); and E. P. Skinner, *Les Mossi de la Haute-Volta* (Paris: Les Éditions Internationales, 1964).

[3] On political invention in traditional Mossi society see C. Savonnet-Guyot, 'Le Prince et le Naaba', *Politique africaine*, 20 (Dec. 1985), 29–44.

[4] This relation of sovereignty was in fact extremely variable and depended greatly on the military capacities of the Mogho Naaba of the day.

emperor, ensured that the central authority was respected by all members of the empire's administration.

Being highly decentralized, the Mossi empire owed its cohesion to a wide network of kinship relations, based upon vigorous exogamy—both male and female—and mediated through a complex system of ritual practices.[5] The family was the basic social unit, sharing direct descent from a common founding ancestor. From the centre to the periphery Mossi society was woven into a system of exchange and interaction which relied on total adherence to the group and its norms. There was thus complete individual and collective integration into the socio-political system.[6]

Mossi society with its tight organization, based upon an extreme religious cosmogony, resisted all foreign influence and preserved its integrity for centuries. Thus the Mogho Naaba's empire represented 'the unique example of a Sudanese group which had preserved its ethnic identity and political autonomy through all the vicissitudes of Sudanic history until French conquest in 1896'.[7] It should be noted that all pre-existing societies in the territory of modern Burkina (the Mossi kingdoms, but also the acephalous 'stateless' groups in the west and south, such as the Bobo and the Dagari) were unwilling to accept Islam. The Mossi in particular refused the social and political order of conquering Muslims like Sonni Ali or Askya Muhammad, considering it incompatible with the cult of ancestors upon which their state apparatus was based. Moreover, right through the Middle Ages their history was one of endless conflicts with neighbouring Islamic groups (Timbuktu was sacked by the Yatenga Mossi in 1328 and 1333, and Sonni

[5] See R. Pagard, 'Les Relations de la descendance avec les aïeuls et les bisaïeuls chez les Bambara, les Mossi et les Peul: Contribution à une étude comparée de la famille dans la savane de l'Ouest-africain', *Notes et documents voltaïques*, 5/4 (July–Sept. 1972), 49–54.

[6] On the socio-political organization of the Mossi empire cf. Izard, *Introduction à l'histoire des royaumes Mossi*; Skinner, *Les Mossi de la Haute-Volta*; id., 'Processus de l'incorporation politique dans les sociétés africaines traditionnelles: Le Cas des Mossi', *Notes et documents voltaïques*, 1/4 (July–Sept. 1968), 29–47; A. S. Balima, *Genèse de la Haute-Volta* (Ouagadougou: Presses Africaines, 1969); B. B. Somé, 'La Religion traditionnelle mossi comme source de valeurs de civilisation politique', *Notes et documents voltaïques*, 4/1 (Oct.–Dec. 1970), 3–31; and P. Ilboudo, 'Croyances et pratiques religieuses traditionnelles des Mossi' in *Recherches voltaïques*, iii (Paris and Ouagadougou: CNRS and CVRS, 1966), 3–32.

[7] E. P. Skinner, 'Christianity and Islam among the Mossi', *American Anthropologist*, 60/1 (Dec. 1958), 1102.

Ali and Askya Muhammad won victories, in 1477 and 1498 respectively, over the Mossi cavalry, previously considered invincible). Despite changing fortunes the Mossi kingdoms fiercely resisted Islam for several centuries, although they were unable to oppose the great *jihād*s in the nineteenth century.

While Islam was therefore introduced into the country of the Mossi relatively recently, Muslim theocracies elsewhere south of the Sahara had officially declared adherence to the Qur'ān several hundred years before. Indeed the arrival-date of the first Muslims can be etablished fairly precisely: it coincided with the arrival of the first Yarse or Mande traders in the seventeenth century.[8] Because of the goods they provided these traders were tolerated by the *nanamse* (plural of *naaba*, 'chief'); but they gradually acquired considerable influence in the courts where they were received. Islam's progress nevertheless continued to be slow, since most of these traders attached scant importance to proselytism. Having been stagnating in the eighteenth and most of the nineteenth centuries, Islam only started to expand during the colonial conquests.

This was when the first Mossi emperor (Mogho Naaba Doulougou) was converted to the faith of the Prophet. It was in his reign that the first mosque in Ouagadougou was built and an official imamate was established in the imperial court. However, both this conversion and ensuing ones were unconnected with any concern to underpin the state power with religious legitimacy, as this was already achieved through the traditional religion. Nor was it based upon personal conviction (except in a few rare cases). Its foremost practical function was to reinforce the loyalty of alien Muslim groups towards the traditional authority. Thus the occasional adoption of Arab names by the Mogho Naabas, as well as their participation in prayers (*ṣalawāt*) or the celebration of Muslim feasts, became rituals through which the sovereign periodically ratified the pacts of allegiance of his Muslim subjects.

[8] M. Izard suggests that the first Yarse traders arrived in Yatenga towards the middle of the 16th c.; see 'Les Yarse et le commerce dans le Yatenga pré-colonial', in C. Meillassoux (ed.), *L'Évolution du commerce africain en Afrique de l'Ouest* (English edn.: *The Development of Indigenous Trade and Markets in West Africa)* (both Oxford: Oxford University Press and International African Institute, 1971), 271. N. Levtzion, 'Outlines of the Development of Islam in Mossi (Wagadugu)', roneo (University of London: SOAS, undated), suggests that the Yarse arrived in the Moaga country in the beginning of the 16th c.

So Burkinabe Islam remained marginal for a long time. This marginality was apparent at two levels. First, it was ethnic in the sense that Islam's original promoters, mainly the Yarse, were ethnic strangers among the Mossi people. Although often tolerated, and sometimes honoured, the Yarse were invariably kept in a position of total subordination. True, they had certain privileges (for instance their monopoly over the imamate attached to the royal court) which gave them a disposable income in addition to the magic aura of their religion; but these were personal arrangements which never called into question their total dependence upon the Mogho Naaba. Secondly, Burkinabe Islam was marginal in a social sense because the Muslims constituted only a negligible minority of the Mossi people. They were free to practice their religion, but within very strict limits dictated by respect for the traditional order.[9] The noble castes rejected Islam because it condemned the cult of ancestors which was the symbolic cornerstone of their power. Lower social orders dismissed it because pronouncing the *shahāda* was tantamount to a complete rupture with their group. Muslims often had to hide to carry out their ritual *ṣalawāt* as they were prohibited from all public demonstration of their faith. They rarely got permission to construct mosques or Qur'anic schools. The *imām* at the court of Ouagadougou—the institution was established at the kingdom of Yatenga several decades later—had to swear, by God and the *Nabiyama* (the Prophet), his loyalty and submission to the Mogho Naaba, who invested him with the symbols of his function.[10]

It was logical for Islam to remain politically weak in Burkina as long as it was socially marginal. Indeed, the Mossi sovereigns were at pains to preserve their political system from all Islamic influence. Although converted, the Mogho Naaba usually prohibited his eldest son from attending one of the few existing authorized Qur'anic schools, thus forestalling a possible future break with tradition. The Muslim Fulbe of Yatenga were excluded from Moaga institutions because they remained outside the system of land consecration.[11] The kings and several

[9] On the organization of the Yarse in Mossi country see A. Kouanda, 'Les Yarse: Fonction commerciale, religieuse et légitimité culturelle dans le pays moaga', *Doctorat de troisième cycle* (UER d'Histoire, Université de Paris I, Panthéon Sorbonne, 1984).

[10] A turban and a white garment; see Skinner, *Les Mossi de la Haute-Volta*, p. 298.

[11] See M. Izard, 'La Politique extérieure d'un royaume africain: Le Yatenga au XIX<sup>e</sup> siècle', *Cahiers d'études africaines*, 3–4, 87–8 (1982), 368.

members of the royal family in Yatenga remain faithful to their traditional religion *to this very day*, even though most of their subjects are now converted to Islam.[12] On the whole it seems that the rare conversions which occurred in precolonial days involved only people 'who did not hold strategic positions *vis-à-vis* the ancestors'.[13] In Yarse villages, which enjoyed a wide measure of local autonomy, the Mogho Naaba continued to appoint the *imām* and the Yar Naaba (the Yarse chief), the former being formally subordinate to the latter.[14] Perceived as a means of emancipation from the ancestral hierarchies, Islam remained subject to a number of material and symbolic constraints which prolonged its stagnation until the traditional system of inequality and domination was destroyed.

### *A Branch of Islam Cut off from the* Umma

The Mossi kingdoms' impermeability to outside influences, maintained through endless conflicts with neighbouring Islamic states, profoundly influenced the relations between the local Muslims and the rest of the *umma*. Spiritually the Mogho and the Qur'ān were completely incompatible.[15] Confronted with the choice of integrating, and preserving their rights of citizenship, or preserving all the distinctive features of their separate identity at the risk of being rejected by the Mossi people, the Muslims opted for the former solution, even though this implied depriving their religious life of much of its substance.[16] In this connection, E. P. Skinner notes that 'despite conversion to Islam, the Muslim Mossi have retained the major values of their traditional culture'.[17] Islam was thus entirely annexed by the 'national' culture, but deprived of universalist transcendence. Contact with the *umma* was not entirely broken off; but the Yarse, who travelled widely to carry on their trade, appeared

[12] See Izard, 'La Politique extérieure du Yatenga', p. 382.

[13] Skinner, 'Christianity and Islam', p. 1106.

[14] See Kouanda, 'Les Yarse', p. 165.

[15] T. F. Pacere, *Ainsi on a assassiné tous les Mossé* (Sherbrooke, Quebec: Editions Naaman, 1979), 45.

[16] See Pacere, ibid.

[17] E. P. Skinner, 'The Diffusion of Islam in an African Society', *Annals of the New York Academy of Science*, 96/2 (Jan. 1962), 666.

more preoccupied with gold, salt, and kola nuts than with enriching their faith and propagating the Qur'anic precepts.

Attentive reading of the colonial literature (official reports, studies, and other university research from the 1950s) on Burkina's Muslims reveals a constant preoccupation with the low general level of education of Islamic agents (*marabouts*, *imāms*, and Qur'anic teachers) in the Moaga country. But naturally conditions varied, especially according to ethnic groupings. The Toroobe in Yatenga appear to have been a true *marabout* clan who mastered Islamic scholarship and practised Sufism and the cult of *awliyā'*; but their influence was limited because of their weakness as proselytes.[18] All the colonial writers displayed a distressing tendency systematically to underestimate the progress of Islam among Mossi animist populations. They tried eagerly to emphasize the 'lukewarm' aspect of Bukinabe Islam (especially among the Mossi), which they contrasted with the Islamic revival elsewhere in West Africa, thus bearing witness to the colonial administration's mistrust of the pan-Islamic exuberance of the more violently anti-European movements (Wahhabis, Senoussis, and Mahdists).[19] Nevertheless, the corroboration of their judgements by those of researchers who reject the anti-Islamic ideology of certain colonial administrators indicates that their account is not only plausible but also accurate.

Islam expanded in West Africa through continued links between African Muslims. Djenne, Timbuktu, and Kankan (not to mention Tlemcen, Al Azhar, Medina, and Mecca) had Islamic universities where West African Muslims gathered for study and spiritual guidance. The *mallam* (*more* in Moore, the language of the Mossi) who wished to improve his scholarship continued his journey in search of the most famous teacher until he felt he had found him. Until the beginning of the twentieth century and the arrival of the modern communications needed for the development of the colonies, contacts between Muslims

[18] See H. Diallo, 'Notes sur l'islam au Yatenga jusqu'à la fin du XIX<sup></sup>e siècle', unpublished dissertation.

[19] An attitude which was often linked with a certain contempt for 'uncultured' or 'ignorant' *marabouts*, to use B. Bichon's terms, from 'Les Musulmans de la subdivision de Kombissiri (Haute-Volta)', in *Notes et études sur l'Islam en Afrique Noire* (Paris: Recherches et Documents du CHEAM, 1, 1962), 85.

from the Moaga country and the great West African Islamic centres were rare. Few Qur'anic teachers learnt Arabic, the language of the Qur'ān, or mastered Islamic sciences (such as *fiqh*); and few Qur'anic schools had more than ten *ṭalibs*. (In certain cases one can indeed identify an erudite *marabout*.[20] But he is usually a contemporary figure who has benefited from the indirect boost to Islamic development provided by the decline of traditional Mossi society. His fame has rarely crossed the boundaries of his region of origin, and the reverence accorded him tends to be limited to an intimate circle of the faithful.) The initial handicap of cultural isolation was further aggravated by the rules which governed appointments to the imamate: for an applicant to be successful, it was necessary (but also sufficient) that he was of Yarga (singular of Yarse) origin. Neither literary nor intellectual competence was taken into account.[21] In short, without access to sources of Islamic scholarship and in isolation from the *umma*, Islam has had little influence in Burkina and has remained confined to the formal and strictly legal and ritual aspects of the Qur'ān.

Islamization in Africa usually involves brotherhoods. The mastering of mysticism has turned these virtually into enterprises for the institutionalized management of charisma. It is through the brotherhood that charisma has prospered and produced *awliyā'*. But the brotherhoods did not penetrate Burkina until the beginning of the twentieth century.[22] It is extraordinary that no real study, even from a partial viewpoint, of the brotherhood movements in Burkina has yet attempted to

[20] There are references to learned men famous for their erudition in M. Max, 'Notes sur l'islam et le christianisme dans la subdivision centrale de Ouagadougou', mémoire, CHEAM (1956); E. P. Skinner, 'Islam in Mossi Society', in I. M. Lewis (ed.), *Islam in Tropical Africa* (2nd edn.; London: International African Institute and Hutchinson University Library for Africa, 1980), 187; and Bichon, 'Les Musulmans de la subdivision, p. 97. The most remarkable example is that of Souleymane Kafande, a Qādiri teacher regarded as a *walī* and miracle worker, whose fame reached Bawku in Ghana and who is mentioned by both Bichon and Skinner.

[21] See M. Campaoré, 'L'Islam en Haute-Volta', *Mémoire de DEA* (Université de Dakar, Faculté des Lettres et Sciences Humaines, 1983/4), 13. Kouanda, 'Les Yarse', p. 165, suggests that this rule originated in the Mossi sovereign's concern to turn the Yarse into a tolerant counterweight to the rigorous Islam preached by 'Uthman b. Fudi's *jihād*.

[22] According to J. Audouin and R. Deniel, in *L'Islam en Haute-Volta à l'époque précoloniale* (Abidjan: INADES, 1975), 41, the Qādiriyya was introduced into Ouagadougou in 1920.

explain this phenomenon, unique in West African Islam. Whatever the reason, the *ṭuruq* (plural of *ṭarīqa*, 'spiritual way') did not, and do not even now, constitute structures through which Muslims could have created, or could create, a counter-movement around an uncontested *shaikh* invested with *baraka*. No Burkinabe refers to himself as Qādiri or Tijāni in the manner of the Mourides in Senegal. Although there are members of all West African brotherhoods—with a clear preponderance of al-Hajj ʿUmar's Tijānis—such adherence has remained formal and has hardly led to a development of mysticism. The only exceptions are some Hamallist centres in the northern region, and some *marabout* clans who have settled at Bobo-Dioulasso, to whom we shall return.

### *A Legalistic and Bureaucratic Form of Islam, Under Challenge*

An important consequence of the weak impact of Sufism in Burkina was a growing legalistic tendency within Islam; legalism and charisma are hardly synonymous. Before the colonial period, the Muslims' subordinate status had forced them to reduce their religious practices to a purely formal level. At the arrival of colonialism, when the Moaga country was opened up to the rest of West Africa and the *umma*, the legalist trend was strengthened, even though this was also the time when Mahdist and prophetic phenomena began to emerge.

Decolonization in Black Africa led to the grouping of Muslim communities within 'national' boundaries. In Upper Volta this phenomenon resulted in the rise of reformism, a movement which spread throughout West Africa in the 1950s under the auspices of the UCM (Union Culturelle Musulmane). Very soon the success of Muslim reformism gave rise to the establishment of the CMHV (Communauté Musulmane de Haute-Volta, now Communauté Musulmane du Burkina Faso). Today the CMBF is the most highly organized Islamic movement south of the Sahara. It is extremely hierarchical and centralized, covering the country with a network of local and regional committees responsible to a National Council which has set up several special commissions.[23] The institutionalization

[23] On Teaching and education, Organization, Information and propaganda, Youth, Pilgrimage, and Property rights.

of Burkinabe Islam did not obviously imply a split between reformists and traditionalists, as was the case elsewhere where reformists overtly opposed 'maraboutic power' on ideological grounds. But the CMBF was distinctly hostile towards the 'superstition' and 'charlatanism' of the *marabouts*. The 'modernist' outlook entailed deep mistrust of any emotional demonstration of mysticism or Sufism, and therefore of the cult of saints, as this was by definition uncontrollable.

In contrast to many similar sub-Saharan organizations the CMBF started out with a large measure of autonomy. But because of these religious divisions it proved unable to prevent the post-colonial state from exercising broad control over it. It has few members among the modern élites or the ruling class; its political resources are limited; and it is handicapped by traditions ill-suited to the demands of a modern state. As a result the Muslims have been unable to avoid a total bureaucratization of their main representative institutions. The CMBF has become part of the bureaucratic machine and the political powers use it as a protective instrument of social control against the emotional excesses to which charismatic phenomena are inclined.[24]

Burkinabe Islam is not only legalistic and bureaucratic; it is also under challenge. First, there is competition within Islam: the leaders of the CMBF have been fiercely challenged by members of the Wahhabiya. CMBF officials have had to harden their attitude towards any religious practice which might be considered heterodox, especially if it might be exploited by the Wahhabis (whose success is based on their legitimate links with the *umma* and whose propaganda stresses pure faith and respect for dogma while condemning deviations).[25] There is also competition from a dynamic and influential Catholic church with a substantial social infrastructure. Ever since independence the vast majority of the political élite in charge of the post-

[24] On CMHV, see R. Otayek, 'La Crise de la communauté musulmane de Haute-Volta: L'Islam voltaïque entre réformisme et tradition, autonomie et subordination', *Cahiers d'études africaines*, 24/3/95 (1984), 299–320.

[25] The Wahhabis in Burkina belong to the *Jamia'at Ahl Al Sunnā Al Muhammadiā*; this name stresses the Wahhabis' respect for orthodoxy and the intangibility of dogma. See Otayek, 'La Crise de la communauté musulmane', p. 310–11.

colonial state have been from Catholic schools, long consi... a weapon in the battle against Islam.[26] Muslims were deprived of a modern education, which in any case they considered liable to subvert traditional morality. But Qur'anic instruction did not offer them the kind of skills they needed to challenge the hegemony of the élites with their 'European' education. The first *madrasas* adapted to the demands of a modern state only appeared in the 1950s, in the face of hostility from the *marabouts*, who considered them a threat to their influence, even to their earning power. Despite its faults (for instance the Church's unequivocal support for the discredited Yameogo government of 1960–6) Christianity was seen by many Muslims to provide the best opportunities in the modern world.[27] Aware of this situation, and hence of the attraction for young Muslims of a Christian education, Burkina's Islamic leaders recognized the need for greater pragmatism and adjusted their discourse accordingly, even though such westernization aroused identical criticism from both the Wahhabis and the traditionalists headed by the *marabouts*.

Owing to these particular factors the Islamic model in Burkina is different from that of any other West African country. They explain why charisma is so relatively insignificant in this model. The process of Islamization in Burkina had only few points in common with what went on elsewhere in West Africa, even if certain major developments occurred everywhere. Nevertheless, though there have been relativelay few cases of Muslim charisma in Burkina Faso, it would be a mistake to overlook the continued existence of phenomena which are basically charismatic, though less intense than elsewhere and of a rather peculiar nature.

[26] 'We need schools, cost what they may . . . We must guide the élite which will become tomorrow's ruling class.' Mgr. Thévenoud, former apostolic vicar in Upper Volta, in *Vivante Afrique*, 223 (Nov.–Dec. 1962), 29; see also B. Nouaille-Degorce, 'Les Pères blancs en Haute-Volta', *Mémoire, diplôme supérieur de recherches et d'études politiques* (Institut d'Études Politiques, Bordeaux, 1973), 77 ff.; and D. Bouche, 'Les Écoles françaises à l'époque de la conquête, 1884–1900', *Cahiers d'études africaines*, 6/2/22 (1966), 257–60.

[27] On Muslim attitudes to education see R. Deniel, *Croyances religieuses et vie quotidienne: Islam et christianisme à Ouagadougou* (Paris and Ouagadougou: CNRS and CVRS, 1970), 186–98.

## 2. CHARISMA . . . AFTER ALL

Until the colonial period Burkina and its Muslims were isolated and on the periphery of the political upheavals, philosophical currents, and reforms which characterized neighbouring Islamic countries. The arrival of European colonialism tolled the knell of the Mogho Naaba's Empire and provided entirely new opportunities for the development of Islam.

### *Crisis and Charisma*

The political and administrative decentralization which had held the Mossi empire together became a factor of destabilization towards the end of the nineteenth century. Conflicts between the kingdoms and succession quarrels within them were intensified through foreign intervention. Less than fifteen years after the signing of the Berlin Treaty (1895) the whole territory was under French rule, though its borders were not drawn up until 1947.[28] The administrative units of the Moaga country were completely changed. The Mogho Naaba was stripped of his judicial powers and the chiefs integrated into the colonial administration as tax-collectors. The foundations of Mossi society were threatened.

As traditional authority weakened, and customary means of social control broke down, Islam appeared to be the religious force of resistance against the European invasion. In the words of M. Gilsenan, the *baraka* of the *marabouts*, often perceived as the soul of Islam, provided a 'refuge' for the people neglected by colonization.[29] Although endowed with exceptional powers, either as 'invulnerable to gunfire' or as workers of miracles, these *marabouts* became known as 'leaders of men' rather than for their charisma. Their main source of strength was that they remained outside both the opposing systems, traditional and European.

[28] In fact the political and administrative status of what is now Burkina was changed several times, according to the whims of French colonial policy: 1896–1904: the establishment of a territory under military administration; 1904–18: attachment to the colony of Upper Senegal and Niger; 1919–32: reconstitution of the territory under the name of 'the colony of Upper Volta'; 1932–47: division of the colony into the Ivory Coast, the Sudan, and Niger; 1947–60: reconstitution of the colony of Upper Volta within its present borders.

[29] M. Gilsenan, *Recognizing Islam* (London: Croom Helm, 1982), 15.

But there were relatively few such *marabouts*, and Islam was a less significant force in the Moaga country than in the rest of Upper Senegal and Niger. Moreover, it is interesting to note that whereas most Muslims considered hostility towards the colonial power a means to overcome their social inferiority, through the safeguarding of Moaga traditions, the *marabouts* saw it as a way of substituting their authority for that of chiefs who collaborated with the French. Islamic charisma occurred in conditions of political and social crisis; but it lacked a vital ingredient, Islamic knowledge, and its significance was bound to decrease as soon as the Moaga country had been 'pacified'. This is indeed what happened.[30]

Charismatic phenomena tended in pre-colonial Burkina to occur in conflict situations. At Bobo-Dioulasso, an Islamic centre in Burkina, the Saghanogho lineage (originally from Kong in what is now the Ivory Coast) succeeded in imposing their influence and status as a holy *Karamogow* (*marabout*) clan by mediating in a late nineteenth-century conflict between the Bobo Dioula and the Ouattara, the Dioula warrior clan.

Like the Kunta of Timbuktu the Saghanagho derived substantial revenues from their special relations with the caravans which traded in gold and salt. Like other *marabout* clans before them (such as the Kassamba, the Fofana, the Cisse, and the Toure) they acquired distinction and wealth thanks to their magical powers. Each Sonangui lineage (a politico-military clan) was served by a *Karamagow* lineage supposed to provide divine protection for its continued authority;[31] but none had succeeded in exercising any influence over the animist (or formally Islamized) people of Bobo-Dioulasso.[32] Like the *Karamogow* clans, the Saghanogho derived their spiritual strength from Islamic knowledge and *baraka*, conferred upon them by this knowledge in the eyes of non-Muslims. But the

[30] A useful analysis of the correlation between charisma and social crisis is to be found in R. H. Dekmejian and M. J. Wyszomirski, 'Charismatic Leadership in Islam: The Mahadi of the Sudan', *Comparative Studies in Society and History*, 14/2 (Mar. 1972), 193–214.

[31] See L. Quimby, 'Islam, Sex Roles and Modernization in Bobo-Dioulasso', in B. J. Rosette (ed.), *The New Religions of Africa* (Norwood, NJ: Ablex Publishing Corp., 1977), 204–5.

[32] See B. Traoré, 'Le Processus d'islamisation à Bobo-Dioulasso jusqu'à la fin du XIX^e^ siècle: Approche historique et sociologique', *Mémoire de maîtrise* (Université de Ouagadougou, 1984), 86.

reason why they, unlike any clan before them, eventually succeeded in spreading Islam in Bobo-Dioulasso was their role as mediators, similar to that described in Gellner's brilliant study of the saints of the Upper Atlas in central Morocco.[33] They adopted this mediating role at a particularly troubled period, when the traditional political hegemony in Bobo-Dioulasso was threatened by the emergence of new influences.

In the extreme north-east of Burkina another case of charisma occurred in the 1930s, during a critical period among the Jelgoobe Fulani in the Djibo region. It was largely because he spoke out against the traditional religious and political authorities, who had supported the colonial administration, that Abdoulaye Doukouré succeeded in establishing around his mosque a following of families who had suffered under French rule. Abdoulaye, a disciple of Shaikh Hamallah and described as an authentically charismatic person,[34] was a famous orator and a model of anti-conformism, a hallmark of charisma.[35] By combining exemplary behaviour with a potentially subversive political attitude Doukouré encouraged the spread of Hamallism in Yatenga, where the doctrine of 'the eleven beads' had already been introduced by Boubakar Sawadogo, another disciple of Shaikh Hamallah. Doukouré and Sawadogo inspired other *shaikh*s who in turn created Hamallist communities.[36] However, though their fame spread throughout Burkina (as is shown by the success of the annual celebration of the *mouloud* (the birth of the Prophet Muhammad) at Ramatoulaye, an important Hamallist centre for the whole of West Africa), Doukouré's successors appear not to have inherited his aura; the *baraka* of his movement has become increasingly routinized and institutionalized.[37]

[33] E. Gellner, 'Pouvoir politique et fonction religieuse dans l'Islam marocain', *Annales: Économies, Sociétés, Civilisations*, 3 (May–June 1970), 699–713.

[34] See P. Riesman, *Société et liberté chez les Peuls Djelgôbé de Haute-Volta: Essai d'anthropologie introspective* (Paris and La Haye: Mouton, 1974); English edn.: *Freedom in Fulani Social Life* (Chicago and London: University of Chicago Press, 1977).

[35] Riesman, *Société et liberté*, 101–2, reports that Doukouré made it his duty to pound millet and carry water, tasks which were always performed by women.

[36] The fact that this doctrine was distorted by the Hamallis, as M. Cardaire confirms in *Islam et le terroir africain* (Koulaba: IFAN, Imprimerie du Gouvernement, 1954), 153, is only of relative importance here and in no way calls Doukouré's role into question.

[37] On the Ramatoulaye pilgrimage see S. Zerbo, 'Fête du Mouloud: Les Pélerins au sanctuaire du Ramatoulaye', *Carrefour africain* (Ouagadougou), 810 (23 December 1983).

The last case I shall quote is that of al-Kari de Bousse, in the Dafing region in the north-west of present-day Burkina. Al-Kari, whose real name was Amadu Deme, was an intellectual educated at Djenne. When he returned from Mecca in 1880 he launched a *jihād* against the Samo animists and also against the plundering Muslim Fulbe of Barani and Toucouleur of Macina. After a time of changing fortunes he died in battle against the French in 1884. So al-Kari de Bousse's *jihād*, a very rare example of holy war in Burkina, developed out of three different conflicts, involving the Muslim Markas and these three groups of adversaries. In the circumstances it is unlikely that al-Kari acquired his reputation because of extraordinary powers or mysticism, even if he is alleged to have belonged first to the Qādiriyya and then to al-Hajj 'Umar's Tijāniyya.[38] If his name is significant (al-Kari is a derivation of the Arabic *al-qāḍī* which means judge or lawyer), it seems that Amadu Deme was an austere and rigid dogmatic rather than an esoteric Sufi—a popular leader rather than a charismatic figure.[39]

*Prophetic Charisma*

If the prophet (in the Weberian sense) and the *walī* are two slightly different forms of the charismatic type, then it must be recognized that it was especially the former that appeared—and appears—in Burkina. No research has shown the *walī* to be a common fellow-traveller of Burkina's Muslims. He has rarely figured either in collective memory or in oral tradition. In contrast, the prophet is a fairly familiar figure in contemporary Burkinabe Islam. And it was prophets who created the most violent opposition movements against the colonial presence during the first half of this century.

This fact is easily explained by the low level of Islamization in Burkina at the time. The brotherhoods were non-existent or just appearing. The complexities of Islamic knowledge were the prerogative of a few intellectuals. Excluded from the élitist system both by tradition and because of socio-political changes introduced by colonialism, the masses created their own men of

[38] See M. J. Echenberg, 'La Jihad d'Al Kari de Bousse: Un état marka en pays dafing à la fin du XIX^e^ siècle', *Notes et documents voltaïques*, 3 (Apr.–June 1970), 3–43.

[39] On this distinction see Dekmejian and Wyszomirski, 'Charismatic Leadership in Islam', p. 199.

authority, able to express (more or less coherently) their frustrations and ambitions. In contrast to a *walī* whose language would be unintelligible in such a context, the prophet's discourse was simple, corresponding precisely to the demands of the situation and concentrating on the rejection through *jihād* of the *Naṣrānī* conqueror (literally, one from Nazareth; by extension, European and white).

Islam was not always behind revolts, which were often sparked off by a refusal to do forced labour or to join the army. Muslims had been known to join movements started by animists, thus giving them—a posteriori—religious legitimacy.[40] Nevertheless, the success of Mahdist ideas is irrefutable, as is shown by the increase in messianic movements during the first half of the century. Al-Kari de Bousse, mentioned above, was converted to Mahdism in the Sudan on his way back from Mecca.[41] Mahdist causes began to flourish in Dori in north-eastern Burkina about 1912, after a circular from Governor-General W. Ponty had discussed the surveillance of Muslim activities. Moussa Aminu, an exceptional Hamallist *marabout* according to R. L. Moreau,[42] proclaimed himself 'saint of his era' and later, when the administration's interference increased, Mahdī, whereupon he was arrested and imprisoned. The colonial administration had to cope with many similar movements, but they were usually localized and quickly suppressed.[43]

However, one such movement flared into a minor but violent revolt, in Bobo-Dioulasso in 1941. On 3 August that year a group of Muslims, led by a *marabout* called Ousmane Traoré, described in official reports as a 'mystic madman', attacked European residents in the town, killing five and wounding nine. An enquiry showed that the revolt was inspired by a letter from Mecca—the famous letter which was circulating in French West Africa at the time—which predicted 'important events' in

[40] For an illustration of this see B. Gnankambary, 'La Révolte bobo de 1916 dans le cercle de Dedougou', *Notes et documents voltaïques*, 11/3–4 (Apr.–Sept. 1978), 1–38.

[41] See Echenberg, 'Le Jihad d'Al Kari', pp. 14–15.

[42] R. L. Moreau, 'Les Marabouts de Dori', *Archives de sociologie religieuse*, 17 (Jan.–June 1964), 127–8.

[43] This is mentioned, *inter alios*, in Balima, *Genèse de la Haute-Volta*, p. 75; Skinner, 'Islam in Mossi Society', p. 179; J. Y. Marchal, *Chronique d'un cercle de l'AOF: Recueil d'archives du poste de Ouahigouya (Haute-Volta) 1918–1941* (Paris: ORSTOM, 1980), 23 n. 3.

1942.[44] It should be noted that the colonial administration helped 'inflate' such events by adopting repressive measures in order to set an example and stop the Muslim unrest which was emerging throughout West Africa, considered all the more dangerous as it seemed to be inspired by pan-Islamic nationalist movements in the Middle East and North Africa. But the revolt is significant for our analysis of prophetism. As Diawara shows, all the rebels of 1941 came from 'the lower classes' that is from groups excluded from the new colonial order. And none of the *marabouts* who led the demonstrations against the colonial power came from the big *Karamogow* or Bobo-Dioulasso families, who never became involved in the revolts.[45] As Weber stressed, prophetic religion coincides with social marginality.[46] The prophet was often an outsider—through force of circumstances rather than by choice—as were his followers.

## *A 'Peripheral' Charisma*

In Burkina such marginality was not only social but ethnic and geographical as well. The *marabout* agitators had shared the marginal status of all Muslims in the Mogho Naaba's empire at the arrival of colonialism. And the revolts were often inspired by *marabouts* of non-Mossi origins from neighbouring colonies, who were thus doubly marginal.[47] Accordingly the most significant charismatic phenomena occurred among non-Mossi groups at the edge of the Mogho. Exceptions were certain acephalous clans and lineages who had resisted Islam and preserved a non-stratified social system which made them hostile to the idea of any individual having abnormal powers over others.

One cause of the spread of charisma at the fringes of the Mogho was the 'Muslim policies' of the Mossi kings, who sent all *marabouts* who wished to create Qur'anic schools of any size to

[44] See Y. Diawara, 'Sur la révolte du 3 août 1941 à Bobo-Dioulasso', *Notes et documents voltaïques*, 8/2 (Jan.–Mar. 1975), 6 ff.

[45] For the Saghanogho, the most influential of these families, this attitude could be justified on doctrinal grounds as a refusal of the *jihād*. See Traoré, 'Le Processus d'islamisation', p. 110.

[46] In J. Séguy, 'Max Weber et la sociologie historique des religions', *Archives de sciences sociales des religions*, 33 (Jan.–June 1972). 84.

[47] Diawara, 'Sur la révolte du 3 août 1941', Diawara, p. 12, and Marchal, *Chronique d'un cercle de l'AOF*, p. 23, both report the case of a *marabout* who came from Salaga (in what is now Ghana) to preach holy war against the Whites in the Moaga country.

the borders of the empire.[48] But the determining factor was undoubtedly that the outlying regions were in contact with peoples of ancient Islamic traditions. Bobo-Dioulasso had links with the kingdom of Kong and with the Islamic centre at Djenne. Ouahigouya was influenced by Macina, and Dori by the Senoussi of Ouaddai, via Sai and Birni, and by the Caliphate of Sokoto. The spread of Muslim 'cultural mores' to these regions was a natural result of their geographical position.

Nevertheless, the phenomenon of charisma took—and takes—different forms in the various regions. In Dori the *jihād* ideal is historically significant in that the Liptako Emirate surrounds a town born of the holy war against the animist Gourmantche; but it is in Bobo-Dioulasso that millenarian ideas have had the most violent consequences. Similarly, though both of these regions have appeared receptive to Wahhabi fundamentalist ideas, the cult of saints has remained important in Bobo-Dioulasso whereas it is non-existent in Dori, as are all forms of local pilgrimage.[49]

If a true charismatic 'centre' exists in Burkina, it is among the Hamallist communities who have settled around Ouahigouya. Significantly, the Jelgoobe Fulani, who make up a number of these communities, were among the first Burkinabes to be Islamized, largely because they lived next to the theocratic Muslim state of Macina. But though the Jelgoobe Fulani proved receptive to Macina's religious influence they resisted all attempts at political tutelage. Indeed they exploited the secular rivalry between Macina and the kingdom of Yatenga to enhance their independence.[50] Hamallism was successful among the Jelgoobe Fulani (and others, notably the Ouahigouya Mossi) precisely because it offered an opportunity to escape political dependence by staying outside or on the fringes of the existing structures at a time when their identity was threatened by the new colonial order. An important factor was the emergence of Abdoulaye Doukouré, whose persuasive discourse suggested 'the way to spiritual, if not temporal, independence'.[51] Henceforth *hijra* (exile or retreat), the Hamallists' ritualized

[48] See Kouanda, 'Les Yarse', pp. 286–7.
[49] See Moreau, 'Les Marabouts de Dori', pp. 118, 134.
[50] See Izard, 'La Politique extérieure du Yatenga', especially pp. 376–7.
[51] Riesman, *Société et liberté*, p. 103.

rejection of the world, acquired a specific location. New economically self-sufficient villages or Hamallist areas were created around such venerated *shaikhs* (as Ramatoulaye, Todiam, You, Taslima, Mansilaye, and Dinalaye).[52] Charisma had been confined to 'the periphery of the periphery'.

## CONCLUSION

The absence of charisma in the pre-colonial and colonial Moaga country, compared with its relative abundance in the outlying areas in the south-west, north-east, and north-west, allows us to make two basic observations about the best conditions for Muslim charisma.

The first is general: charisma is not experienced within societies or groups whose cohesion, not to say unity, is based on an accepted and internalized system of strong social control. In his analysis of Islam in northern Lebanon Gilsenan shows that under such conditions *baraka* is either excluded from the religious and ideological sphere or it is invested in legitimizing the established order.[53] Hidden *baraka*, wholly or partly outside this order, becomes what Séguy has called 'implicit protest'.[54] As it is a sign of latent conflict implicit protest may lead to true charisma where it is expressed by men of exceptional powers. But when tolerated it is usually rendered 'devoid of disruptive potential through a policy of negotiation and subordination'.[55] For the Moaga Muslims implicit protest consisted in ostentatious conduct (habits of dress and food, or refined manners aimed at demonstrating the moral superiority of their religion) which, consciously or not, expressed a certain challenge to the established order. But the challenge did not threaten the equilibrium of this order because the code of Islam implied total submission to the Prince; so Muslim charisma was hidden and circumvented by the principle of *naam*, the divine force which allows one man to command another.[56] Control through *naam* of

[52] See H. Diallo, 'Islam et colonisation au Yatenga (1897–1950): Etude de quelques aspects', *Le Mois en Afrique*, 237–8 (Oct.–Nov. 1985), 42.

[53] Gilsenan, *Recognizing Islam*, 110–14.

[54] J. Séguy, 'La Protestation implicite: Groupes et communautés charismatiques', *Archives de sciences sociales des religions*, 48/2 (Oct.–Dec. 1979), 200.

[55] Ibid.

[56] See Skinner, 'Les Mossi de la Haute-Volta', p. 50.

charismatic potential benefited the Mossi *nanamse* to the detriment of those Muslims who might have claimed that divine recognition without which there can be no charisma.

The second observation is more directly linked with Islam: we believe that the phenomenon of charisma can only occur beyond a certain 'threshold' of Islamization. We agree with Brenner that in West Africa 'the way to the supernatural is through knowledge'.[57] Without this knowledge, without Islamic tradition, without reference to illustrious masters who have contributed to the legal sciences and the establishment of the Qur'ān, without the brotherhoods and the Sufi way, achieved through religious devotion free of the straitjacket of dogma, without the intimate organic link with the *umma* which is the basis for universal Qur'anic preaching, Islam becomes a mere legalistic and formalized system of prescriptions and proscriptions—an 'ordinary' religion, stagnating, secondary, residual, and incapable of competing with traditional African religions, especially where these are foundations of the socio-political order.

These observations may serve as working hypotheses and guidelines for possible research. Pertinent in the context of Burkina, they may, if systematically explored, help explain certain peculiarities of the Islamization process in West Africa and perhaps elsewhere south of the Sahara.

[57] L. Brenner, *Réflexions sur le savoir islamique en Afrique de l'Ouest* (Bordeaux: Institut d'Études Politiques, CEAN, 1985).

# 6

# Women, Islam, and *Baraka*

CHRISTIAN COULON

## INTRODUCTION

WOMEN's presence and position in Islam is a theme which has been largely neglected in contemporary research. Women have been presented as absent or marginal within Islam. Women's studies, including research in Islamic societies, have mostly been developed in the last ten years. However, there seems to be an automatic assumption that Islam hardly penetrates into the world of women.

At a pinch it is suggested that the religious world of the female is characterized by its roots in the pagan pre-Islamic world. In a rare study of Muslim women in Black Africa (on the Manga in Niger) M. Bovin attempts to show that women are 'on the periphery of the periphery' in the Muslim world, and that they are the 'mediators and maintainers' of a pre-Islamic culture upon which they depend to retain their autonomy in relations with the men. This characteristic shows itself in family relationships (the importance of matrilineage), in language (the existence of a women's language influenced less by Arabic and Islam than that of men), and in beliefs and religious practices (the women are the guardians of totemism).[1] Most literature tends similarly to stress women's involvement in the cults and rites of possession (*ndöp* among the Lebou and Wolof, *bori* among the Hausa, *zar* in the Sudan, Ethiopia, and Somalia, and *pepo* among the Swahili). This appears to be a reaction to the domination imposed upon women by male Islamic ideology, or even, to use I. M. Lewis's expression, a 'thinly diguised protest movement' against masculine superiority.[2] In short, the penetration and progress of Islam in African societies appears to have

[1] M. Bovin, 'Muslim Women in the Periphery: The West African Sahel', in B. Utas (ed.), *Women in Islamic Societies* (London and Malmö: Scandinavian Institute of Asian Studies, 1983), 66–103.

[2] I. M. Lewis, *Les Religions de l'extase* (Paris: PUF, 1977); English edition: *Ecstatic Religion* (Harmondsworth: Penguin, 1971).

led to a split between two worlds: the Muslim world under male control and the pre-Islamic, 'traditional' world preserved by the women. Thus the women's universe emerges as a sort of cultural enclave surrounded by the world of Islam. Within it their only option is to attempt silently to subvert, to their advantage, the Islamic rules which keep them in an inferior position. This is how the recluse women of Kano succeed, through the intermediation of their children, in carrying on a petty trade that might at first sight seem prohibited by the rules of purdah.[3]

The research of Islamist scholars points in the same direction. Islam is presented as a religion for men, who are its propagators, guardians, organizers, and ardent defenders while women are on the margins of, or outside, Muslim society. Once more we meet the religious division of labour mentioned above. As J. S. Trimingham, one of the more prolific specialists in Black African Islam, writes: 'the men being Muslims and the women pagans . . . This is found everywhere, not merely among the newly converted. More correctly, one should say, men are more deeply involved in Islam whereas women's participation is marginal, for Islam provided them with little scope for ritual participation.'[4] On the other hand he too stresses the female nature of the cults of possession which are so frequent in Islamic societies. J. Boyd and D. M. Last, experienced researchers in African Islam, note 'how very little is actually on record, after so many years of Islamic research in West Africa, either about specific women *shaikh*s or about the organisation of women's Islam in the history of any particular community'. Indeed, as they point out, such subjects are referred to only in passing or as footnotes.[5]

These omissions are not without explanation. Women's studies have been undertaken mostly by women, often feminists, apt to be suspicious towards Islam, which they consider a sexist tool of oppression, and hence inclined to stress the peculiarities and specific characteristics of African women who have

[3] See for instance E. Schildkrout, 'Dependence and Autonomy: The Economic Activities of Secluded Hausa Women in Kano, Nigeria', in E. G. Bay (ed.), *Women and Work in Africa* (Boulder, Colo.: Westview Press, 1982), 429–50.

[4] J. S. Trimingham, *The Influence of Islam upon Africa* (London and New York: Longman/Librairie du Liban, 1980), 47.

[5] J. Boyd and D. M. Last, 'The Role of Women as *agents religieux* in Sokoto', *Canadian Journal of African Studies*, 19/2 (1985), 283–300.

apparently remained outside Islam. For their part scholars of Islam have often internalized the normative and legalistic universe of Islam, which fails to take account of religious conduct at variance with orthodox thinking. Anything of this kind is relegated to popular religion: 'Islam of the people, of women, fools, pilgrims, preachers . . . in a word, religious life'.[6]

Let me make myself clear. I am not denying that Islam tends to marginalize the public role of women in religious life, nor am I refuting analyses which show female religious peculiarities to exist outside Islam and to be strongly influenced by pre-Islamic cultures. I merely want to show that despite their inferior position in African Islam women are not entirely excluded from the religion. Their Islamic practices are not necessarily limited to the domestic sphere, and their fundamental religious beliefs are not based solely on animist traditions. Women participate in their own way in Islam, manipulating it and accommodating it to their needs. They are active Muslims, even if their practices are informal, hidden, parallel, or heterodox; hence it is wrong to relegate the female Muslim universe to this twilight zone where it only appears to belong because of our inability to study it.

Recent studies of Islamic women in the Arab world and, more tentatively, in African societies have provided new insight into the special features of their practices.[7] M. Strobel's study of

[6] B. Étienne, 'Problèmes de la recherche islamologique au Maroc', in Centre de recherche et d'études sur les sociétés méditerranéennes, *Islam et politique au Maghreb* (Paris: Éditions du CNRS, 1981), 202.

[7] On the Arab World see for instance: E. W. Fernea, *Guests of the Sheikh* (New York: Anchor Books, 1985); D. H. Dwyer, 'Women, Sufism and Decision-Making in Moroccan Islam', in L. Beck and N. R. Keddie (eds.), *Women in the Muslim World* (Cambridge, Mass., and London: Harvard University Press, 1980), 585–98; D. Provensal, 'Le Phénomène maraboutique au Maghreb', *Genève–Afrique* 14/1 (1975), 59–78; S. Andezian, 'Pratiques féminines de l'Islam en France', *Archives des sciences sociales des religions*, 55/1 (1983), 53–66.

On Black Africa (in addition to Bovin's study mentioned in note 1, and that of Boyd and Last in note 5) see the following works: M. Strobel, *Muslim Women in Mombasa, 1880–1975* (New Haven, Conn., and London: Yale University Press, 1979); F. Le Guennec-Coppens, *Femmes voilées de Lamu (Kenya)* (Paris: Éditions Recherches sur les Civilisations, 1983); J. Boyd, 'The Contribution of Nana Asma'u Fodio to the Jihadist Movement of Shehu dan Fodio from 1820 to 1855', M. Phil. thesis (North London Polytechnic, 1982).

Apart from these studies, which deal specifically with women in Islam, certain other articles and books also throw light on the role of women in Islamic religious life (many of these are about Islam in East Africa); see for instance: F. Constantin (ed.), *Les Voies de l'Islam en Afrique Orientale* (Paris: Karthala, 1987) notably F. Constantin, 'Le Saint et le

Mombasa women is typical in its indication that women, often excluded from official Islam, tend to seek refuge 'in non-orthodox or popular versions' of the faith.[8] Indeed it is among the Sufi brotherhoods or in popular Sufism that black African women make their presence most strongly felt (though they are not inevitably excluded from reformist movements, some of which envisage an active religious education for women). So it is essentially outside conventional places of worship (such as mosques) that women are active as Muslims. In view of this the notion of *baraka* is crucial: it allows a break from that dichotomy of the public and the private spheres which encloses women in a subordinate universe. Even though women are refused access to the principal functions or public places of worship, nothing prevents them from having *baraka*, worshipping saints, or becoming agents of a particular *marabout*.

Starting from this hypothesis, we shall explore this 'women's Islam' in two stages, first generally and then more specifically. Part 1 aims to provide a broad sociological outline of this 'women's path' based on information gained from written sources or gathered in the field, mostly in Senegal but also in East Africa. Part 2 is an account of a study of a Senegalese female *khalīfa* carried out with O. Reveyrand at Thiès during a field trip in 1984. This case-study seems to open up some promising perspectives for future research.

## 1. THE WOMEN'S PATH

Women's active participation in Muslim religious life has been observed mostly outside the sphere of the legalistic Islam of the *'ulamā'*. Strict segregation of the sexes and condemnation of the worship of saints or *marabouts*, both features of rigid Islamic practice, tend to confine the religious role of women to the

Prince: sur les Fondements de la Dynamique Confrérique en Afrique Orientale (ibid, pp. 85–109) and C. Coulon, 'Vers une Sociologie des Confreries en Afrique Orientale (ibid, pp. 111–133); A. H. Nimtz, *Islam and Politics in East Africa: The Sufi Order in Tanzania* (Minneapolis: University of Minnesota Press, 1980); R. C. Greenstein, '*Shaykhs* and *Ṭarīqas*: The Early Muslim *'Ulamā'* and *Ṭarīqa* Development in Malawi, 1885–1949', History Department Seminar Paper, University of Malawi (1976).

[8] Strobel, *Muslim Women in Mombasa*, p. 78.

domestic sphere and to children's education—though the latter is highly valued by the Islamic authorities. One gets the impression, as R. A. and E. W. Fernea note, that the less strictly the *ʿUlamā*' control society and its religious life the more women have devloped specific religious activities which reflect 'their own needs and concerns'.[9] The fact that fundamentalist Islamic discourse considers strict moral standards for women an essential cure for economic and social ills reveals an institutionalized mistrust of women. Much Islamic literature describes both Arab and African women as potentially dangerous and a source of disorder (*fitna*) and alleges that some are particularly vulnerable to Western habits, associated in purist minds with decadence and perversion.[10]

This strict form of Islam aims at confining women to their homes, thus limiting their religious life to the private sphere. However, the few available studies of feminine religious expression show that they play a vital part in the popular Islam of the *marabouts*. Though rarely active in the mosques, women tend to make pious visits to the *marabouts* and to shrines. A number of specific religious associations, often with a female hierarchy, are open to them. They may worship women saints or *jinn*-women endowed with special powers. The cermonies which they attend (*ḥaḍra*) accord such importance to dancing, singing, music, and trance that they occasionally turn into virtual therapeutic rites. The search for *baraka* is central to these ritual activities; according to certain writers, in the Maghreb women are the most fervent guardians of *baraka*. In 1900 E. Doutté observed that the brotherhoods were especially popular with women;[11] more recently works on Moroccan Islam and two on Algerian Islam reached the same conclusion.[12] The brotherhoods have emerged as a more typically feminine form of religious practice; the Islam of the brotherhoods and *marabouts*

[9] R. A. and E. W. Fernea, 'Variations in Religious Observance Among Islamic Women', in N. R. Keddie (ed.), *Scholars, Saints and Sufis* (Berkeley, London, and Los Angeles: University of California Press, 1972), 401.

[10] On this fear of women see F. Mernissi, *Sexe, idéologie, Islam* (Paris: Éditions Tierce, 1983).

[11] E. Doutté, *Les Marabouts: Notes sur l'Islam maghrebin* (Paris: E. Leroux, 1900).

[12] V. Crapanzano, *The Hamadsha: A Study in Moroccan Ethno-Psychiatry* (Berkeley: University of California Press, 1973); Dwyer, 'Women, Sufism and Decision-Making'; Andezian, 'Pratiques féminines de l'Islam en France'; Provensal, 'Le Phénomène maraboutique au Maghreb'.

has become primarily the religion of the women. This is because by its rational and explanatory nature it is linked, more than either reformist or fundamentalist Islam, to the traditional values transmitted by women. Hence it is better able to tackle their particular problems and to give expression to their own sociability. Moreover, when the brotherhoods, for reasons which will not be examined here, started declining throughout North Africa, women remained the most faithful clients of the saints and the *marabouts*.

In sub-Saharan Africa the 'popular' Islam of the brotherhoods does not appear to be losing ground, though it is often attacked (sometimes violently) by reformist, modernist, or fundamentalist groups. Women here thus do not act as bastions of a 'traditional' Islam that remains dominant while slowly acquiring greater 'orthodoxy'. As in the Arab world, women are often increasingly present in the brotherhoods, pilgrimages, *dhikr* ceremonies, and celebrations of Islamic festivals (which are stigmatized by the strictest Muslims). Far from being excluded or passive, the women are visibly involved in 'popular' Islam; they are much more discreet within so-called reformist groups except where these pursue a systematic policy of promoting Muslim women (as does the UCM, the *Union Culturelle Musulmane*, in Senegal).

However, such progressive policies as that of the UCM are challenged by other movements appealing for an Islamic revival. Fundamentalist Islam, whose devotees are apt to denounce women 'who cause scandal', recruits hardly any black African women, who for their part consider fundamentalism repressive. Young girls, educated in modern schools, are unlikely to be attracted to groups which advocate their seclusion. Nor can more 'traditional' women see what they would gain from being reduced to a subordinate role, since they are accustomed to a certain status and independence of action and thought, derived from ancient beliefs and customs which were tolerated or even appropriated by the *marabouts*. While reformist and fundamentalist Islam is primarily a masculine religion, 'popular' Islam involves both sexes. At the *magal* in Touba, the annual pilgrimage of the Senegalese Mourides, one is struck by the number of women present, both inside and outside the shrine. Like the men they use the occasion to visit their *marabout*, to organize religious chants, and to wander through the holy city in their splendid *boubous*. In East Africa women participate in the *maulid* celebration (the commemoration of the Prophet's birth), which marks the high point of Islamic life on the coast but which is now opposed by young fundamentalist *shaikh*s, particularly because of the sexual promiscuity associated with it. It is worth noting that this criticism is not new; an ancient dispute between 'Old Muslims' (Arabs) and 'African Muslims' concerns the legitimacy of these ritual ceremonies in which women take part.

These striking examples of feminine Islamic religious practice apart, we still need a more thorough examination of the various ways in which women have become involved in the religious life of the *ṭuruq*.

The mystic Islam of the *marabouts* has produced a number of important women. Some are 'friends of God', some invested with special powers and therefore the objects of popular cults; others are exalted because of the prestige of their scholarship or their maraboutic descent and inheritance of *baraka*. As E. E. Gursoy-Naskali explains: 'The mystic approach was open to men as well as to women, while the ranks of orthodox Islam were closed to the latter.'[13] Certain women are remembered in Sufi hagiographies, for example Rabi'a, who is highly praised in Farid-ud-Din-Attar's '*mémorial des saints*'.[14] Others have been active in the life and development of a particular brotherhood.

Until now little has been known about the worship of women saints in Black Africa. As in the Maghreb, they have been

[13] E. E. Gursoy-Naskali, 'Women Mystics in Islam', in Utas (ed.), *Women in Islamic Societies*, p. 241. This view is closely echoed in E. Dermenghem, *Le Culte des saints dans l'Islam maghrebin* (Paris: Gallimard, 1954). However, to put this information in its proper perspective, it is worth noting the important role played by some women in certain contemporary Islamic movements, for example Zainab al Ghazali, leader of the Association of Muslim Women in Egypt, a true 'passionaria' of the Muslim Brothers. See G. Keppel, *Le Prophète et pharaon: Les mouvements islamistes dans l'Égypte comtemporaine* (Paris: La Découverte, 1983); English edition, trans. by J. Rothschild, *The Prophet and Pharaoh: Muslim Extremism in Egypt* (London: Al-Saqi Books, 1985).

[14] Farid-ud-Din-Attar, *Le Mémorial des saints* (Paris: Éditions du Seuil, 1976), 82–100. The author writes: 'If you are asked: "Why do you include Rabi'a among the men of the élite?", the answer is that there is a *hadis* from the Messenger salvation be unto Him! which says: "Do not pay undue attention to a person's exterior, but take note of his good actions and his good intentions"; this means that any woman whose pious practices and devotion are as agreeable to the Lord Almighty as those of the men of the élite should not be referred to as a woman.' (p. 82).

confused with certain spirits—though in cults of possession (such as the Hausa *bori*) many Muslim spirits are indeed female.[15] On the other hand we are beginning to know more about the influence of female élites within several brotherhoods where they have exerted real religious leadership. As certain (alas fragmentary) data indicate, this may be an old tradition, especially in West Africa. C. C. and E. E. Stewart, in their work on Mauritania, note that several female *shaikhs*, famous for their scholarship and *baraka*, have played a significant religious role in northern Senegal.[16] One of them, Khadija of the Ahl-al-Aqil group, is even said to have been the teacher of the *mujāhid* of Fouta-Toro, ʿAbd al-Qadir, leader of the Torodo revolution.[17] Boyd and Last's research on northern Nigeria showed the important role of the literate women of ʿUthman b. Fudi's family (especially Nana Asma'u, his daughter) in the organization and growth of the cult of the *shaikh*. One of their significant findings is that the activities of these female religious élites were neither incidental nor marginal but a rational feature of a religious reform which promoted the education of women and made them moral guardians of the movement, especially by means of poetry and song.[18] As the two authors suggest, this was not an isolated case. In Senegal certain brotherhoods or brotherhood branches have encouraged the employment of women in Muslim education. The best example is probably that of the Ibrahim Niass movement (the Tijāniyya). The female relatives of the *marabout* were first-rate teachers who created a network of Muslim schools reputed to be among the best in the country. Women (like Mariam Niass and Rokhaya Niass were authentic religious leaders and figureheads of their brotherhood. Though at the outset, like ʿUthman b. Fudi's descendants, they had the special advantage of their family's *baraka*, their scholarship and especially their capacity for organization gave them additional operational roles, as excellent religious

[15] See M. Onwujedgwu, 'The Cult of the *Bori* Spirits Among the Hausa', in M. Douglas and P. M. Kaberry, *Man in Africa* (London: Tavistock Publications, 1969), 279–305.

[16] C. C. Stewart, with E. E. Stewart, *Islam and Social Order in Mauritania* (Oxford: Clarendon Press, 1973), 22–4.

[17] See J. R. Willis (ed.), *Studies in West African Islamic History*, i. *The Cultivators of Islam* (London: Frank Cass, 1979), 12.

[18] See Boyd and Last, 'The Role of Women'.

agents and models of good conduct. In the Nkhota-Kota region of Malawi the Qādiriyya was established, between 1920 and 1930, largely through the missionary activities of a woman, Shaikh Mtumwa, who, having received the *ijāza qādir* in Zanzibar, preached to the underprivileged Africans, especially the women, and popularized a movement which called the religious monopoly of the Arabs into question.[19] Great religious prestige, from the association with the *baraka* of their father or son, was attached to the status of mother and daughters of *marabout* saints; not every woman gives birth to a *walī*. This explains why Diarra Bousso in Senegal, the mother of Amadu Bamba (founder of the Mouride brotherhood), herself from a family which claimed descent from the Prophet, is the object of a special cult. And in the same country the daughters of charismatic *marabouts* are joint heirs to their father's *baraka* with their brothers. They often marry *shaikh*s of the brotherhood themselves, with the result that the state of 'grace' remains within the hierarchy of the order. Thus the wives and daughters of the *marabouts* (the *sokhna*) indirectly perpetuate the *baraka*. Even if they have no religious function as such they become exalted through their rank and lineage and play a crucial role in the social life of the brotherhood (for instance during pilgrimages). In recent years the growing number of female brotherhood associations has caused them to become even more involved in organizing and mobilizing women within the *ṭuruq*. Several of these brotherhoods have many women disciples. They not only possess *baraka*, they also consume it; and they perpetuate it. True, the cult of saints, the pious visits to marabouts, and the recital of the *dhikr* are not their exclusive preserve; but it is in this religious space that they move most easily and make their presence felt. Contrary to general belief, they are a vital element in popular religious life.

Women's active participation in Islam has been encouraged by certain religious chiefs, if only, as Boyd and Last note, because they are often scholars with traditional magical and therapeutic powers which might obstruct the penetration and spread of Islam if not employed in its defence and promotion.[20]

[19] On Shaikh Mtumwa see Greenstein, '*Shaykh*s and *Ṭarīqas*', and Constantin, 'Le Saint et le prince'.

[20] Boyd and Last, 'The Role of Women'.

Alternatively women's ancient practices are integrated or reinterpreted within the new religious order. Women, who are often regarded as a threat to society, have to be won over, mastered, and organized. Strict segregation or exclusion of the female sex might result in 'antisocial' conduct. Whatever the reason, certain modernizing *walīs* or *marabouts* have insisted upon the specific need to promote women within the religious order; in many cases, such promotion has been associated with that of subordinate social categories and provoked opposition among the established élites. This happened at Lamu (Kenya) where the religious innovations of Habib Saleh (for instance the introduction of music and female participation in the *maulid*) caused endless quarrels. Limamou Laye in Senegal, the founder of a small Lebou brotherhood (the Layennes), preached the equality before God of everyone, especially promoting the causes of women and young people, whom he involved in various religious ceremonies (a special place was assigned to women in the mosques; and they replied to the men during the religious chants). Similarly the opening-up to women of the Niass branch of the Tijāniyya led to extended missionary activity within the movement. This popularized *fā'ida* (the saving grace bestowed by God upon a *marabout* during a vision) and *tarbiyya* (the Sufi initiation) and was intended to give a religious experience to everyone rather than limit it to a few initiates. Ibrahim Niass took pains never to exclude women from religious life, and used to explain that nothing need prevent a pious man from touching the hand of a woman, something he often did himself (whereas most *marabouts* tend to avoid such contact).

A few brotherhoods have even become the voices of particular subordinate groups. I am thinking particularly of the Husseiniyya in northern Kenya, whose members, according to A. Hjort, tend to be poor immigrants and 'Somali women frustrated in their marriage'.[21] The dominant groups within the local religious hierarchy consider this *ṭarīqa* heretical and a threat to public order.

There is every indication that women's inclusion in the structures of the popular Islam of the brotherhoods is much

[21] A. Hjort, *Savanna Town: Rural Ties and Urban Opportunities in Northern Kenya* (Stockholm: Stockholm Studies in Social Anthropology, 1979), 40–7.

more significant than has hitherto been believed either through lack of interest or scarcity of information. Numerous women are members of *marabout* organizations. Many take the *wird* and recite the *dhikr*. In Senegal I have noticed that it is becoming increasingly common for women to pay allegiance (*jebellu*) directly to a particular *marabout* rather than accept an indirect or passive membership through the intermediation of their husband or father.[22] What is more, the women may choose their *marabout* independently rather than share the religious guide of their father or husband. This autonomous quest for *baraka* may be related to particular feminine strategies, strategies based upon specific personal or practical considerations, as has been shown by D. H. Dwyer's work on women's membership of Moroccan *marabout* orders. She observes that ultimately such decisions depend on the exigencies of the women's daily lives, on their priorities and their availability.[23] It may be noted that certain *marabouts* make efforts to be in the women's 'good books', frequently attend their meetings, willingly discuss subjects concerning them, and generally improve their image among them.

In Senegal, where *shaikh*s from great *marabout* families are growing in number, the search for a new clientele is becoming essential for their legitimacy. It is hardly surprising that the *marabout* class are paying renewed attention to women and the young. The opening-up of the Senegalese brotherhoods to women is accelerated by the growth in the number of female *marabout* associations, especially in towns—though Islamic women's organizations are not a new phenomenon in the Arab world, nor even in Black Africa; an interesting early example is that of the *Yan Taru*, founded by the daughters of ʿUthman b. Fudi at Sokoto. According to Boyd and Last this association played a decisive role in organizing the cult of the *shaikh*, thereby helping to perpetuate his *baraka*.[24]

The female *dāʾira*s in Senegal are of more recent origin. They are groups of women disciples of a particular *marabout*, and they

[22] D. B. Cruise O'Brien, in his study of the Mourides, does however note that only widows and daughters of *marabouts* paid their own *jebellu*; see *The Mourides of Senegal: The Political and Economic Organization of an Islamic Brotherhood* (Oxford: Clarendon Press, 1971), 85–6.

[23] Dwyer, 'Women, Sufism and Decision-Making', p. 587.

[24] Boyd and Last, 'The Role of Women'.

exist among both the Mourides and the Tijānis. Some women's *dā'ira*s are branches of larger associations which include members of both sexes; others are specifically female. They have various functions: basically they constitute a mutual aid or business network, but they also serve as support groups for particular politicians. Thus one of the most powerful women's *dā'iras* in Dakar, the *Dahiratoul Khairy wal baraka*, whose members are *ṭālib*s of the *Tijāni marabout* Shaikh Tidiane Sy, regularly canvass on behalf of one of their members, A. Aram Diene, a well known woman politician and member of the National Assembly. Moreover, as in the case of the *Yan Taru* at Sokoto, the *dā'iras* organize the publicity for particular *shaikh*s and thus serve as religious agencies. To this end, they organize sessions of religious chanting in honour of their *marabout*, make pious visits to him (*ziyara*s), and miss no occasion to sing his praise or acclaim his *baraka*. Finally, the considerable incomes of some of the women, especially those in commerce, represent an important source of finance for the brotherhood chiefs. A Senegalese *shaikh*'s charisma depends not only on his personal qualities and scholarship, but also on his financial means. The women's *dā'iras* have thus become important influences on the religious market of the *marabouts*.[25]

The religious activities we have described do not belong to the private domestic sphere. Religious chants, *ziyara*s, and pilgrimages to holy places are all public expressions of women's religious feeling and are striking in their 'festive' aspect. Women's brotherhoods and *marabout* associations may well have replaced other women's organizations which have fallen into disuse or been prohibited.[26] The varied opportunities for religious expression allow women to leave the peripheral religious space to which Islam tends to relegate them. Women are present and sometimes, as in Senegal, powerful within

[25] The importance of the female *dā'iras* is pointed out by M. C. Diop in his thesis on Mourides in an urban environment: 'Compared to the traditional rural institution of the *dāra*, the *dahira* is significantly different in the sense that it allows the coexistence of the sexes' ('La Confrérie mouride: Organisation politique et mode d'implantation urbaine', *Doctorat de troisième cycle* (Université de Lyon II, 1980), 166).

[26] Nimtz notes for example that in Tanzania women's membership of the Qādiriyya has risen spectacularly since the Government's decree prohibiting female dancing societies; see *Islam and Politics in East Africa*, p. 304.

Islam, though occasionally they may be using religion as a pretext for a general expression of female sociability.[27]

These reflections on the 'women's path' within Islam are fragmentary and incomplete. Nevertheless, they suggest that in Black Africa women are not always passive and dependent second-class Muslims. Although a specific feminine religious practice appears to exist, this has had its influence on the Islam of men, indeed on general Islamic practice. At Sokoto, as in Senegal, female Muslim associations seem to be vital to the organization and development of the *marabout* cult. Let us now go even further and discover what happens to a Muslim community when a woman becomes its *shaikh*.

## 2. THE DAUGHTER OF THE *SHAIKH*

Sokhna Magat Diop is the only Senegalese woman in charge of a Muslim religious community. She even has the title of Shaikh in the Mouride brotherhood, where she leads one of the 'sections' created by her father Abdoulaye Iyakhine, better known as Abdoulaye Niakhite.

The Mouride brotherhood consists of a number of clans or important families founded by the brothers, sons, or great *ṭālibs* of Amadu Bamba. At the head of each such family is a *khalīfa*, who has authority over the *marabouts* and disciples of the brotherhood branch while being directly responsible to the Khalifa-General of the brotherhood, to whom he must give his allegiance (*jebellu*). The succession in each branch follows the same rules which determine the leadership of the brotherhood itself: the founder is generally succeeded by his eldest son; at the death of this first successor the next brother takes over, to be followed by the younger brothers.

Abdoulaye Iyakhine died in 1943 without male heirs, as his only two sons, Shaikh Diop and Killal Diop, had died a few years earlier. In accordance with Abdoulaye's instructions Sokhna Magat Diop, his eldest daughter and child of his union with Sokhna Tabara Cisse, became head of his Mouride branch.

[27] Le Guennec-Coppens, in her work on Lamu (Kenya), clearly shows that female religious celebrations are in effect occasions for a party; see *Femmes voilées de Lamu*, pp. 61–2.

At the time Sokhna Magat Diop was twenty-six and married to a younger brother of Amadu Bamba, Serigne Massamba M'backé.

This 'exceptional' woman (to use the expression of her eldest son, Mama M'Balla M'backé, our main source of information) has been *khalīfa* for over forty years. Her *zāwiya* is at Thiès, in an area named after her father, where he had come to live. Though in many ways a fascinating Mouride figure, she has aroused little interest among the many researchers interested in this *ṭarīqa*. This may be because her community is not very large in terms of numbers of disciples, or else because her case is considered a marginal one of female religious authority without real significance. We think, however, that this case, although unique, may reveal the mechanisms whereby some women penetrate the *marabout* class, apparently to be totally accepted—'like a man'. But we must go further and ask ourselves what is so special about Sokhna Magat Diop's charisma and whether her community, which includes both men and women, has peculiar characteristics as a result of its female leadership.

Sokhna Magat Diop is an unusual *marabout*; but so was her father. The extraordinary destiny of this legendary figure is somehow being continued through his daughter, as though something unique has coloured the religious experience of them both. According to an oral tradition handed down by his descendants, Abdoulaye Iyakhine came from the East via Morocco. Nothing is known of his father and mother or of his siblings. He was commanded by God, in a dream, to go in search of a master named Shaikh Ibra Fall, whom he finally encountered at N'Dande after a long journey.

Shaikh Ibra Fall was an illustrious Mouride figure. Like his future disciple Abdoulaye he had a divine revelation and was ordered to seek out a *marabout* to guide his life. He found him in the person of Amadu Bamba and became not only his 'prophet', as some of his faithful still say, but also his main propagandist. Shaikh Ibra Fall became one of the pivots of the brotherhood organization. But he also founded, within it, the Bāy Fall sect, which took certain specific Mouride traditions (devotion to the *shaikh*, work in place of prayer) to extremes, leading to a dispensation from obligatory religious practices (prayer or the fast of Ramadan). Despite this eccentric behaviour, the hierarchy of the brotherhood considered the Bāy Falls entirely legitimate Mourides.[28]

Abdoulaye Iyakhine became a disciple of Shaikh Ibra Fall, though he also pledged his allegiance to Amadu Bamba himself. But this disciple was something out of the ordinary; his *ṭālib*s considered him the Mouride figure third in importance after Amadu Bamba and Ibra Fall. They even claimed that he was closer to God than Amadu Bamba, the founder of the Mouride movement—who is said to have admitted this himself. He was reputed to have extraordinary powers: he identified and fought sorcerers (*domm*), he cured snakebites, and he could destroy by fire, from a distance, evil objects or people. A brother of Amadu Bamba who was in debt to him and had insulted him had to face the consequences when his house was mysteriously burnt down; Amadu Bamba had to reconcile them, and urged the Mouride community to accord this powerful personality the respect due to him. The colonial administrators were unable to take his photograph for their card-index of *marabouts*.

In 1930, the year of the death of his master Ibra Fall, Abdoulaye Iyakhine is said to have had himself declared Mahdī and to have proclaimed his intention of supplanting the Governor of French West Africa, intending then to share out the services of the French administration among his disciples. His descendants now reject this allegation, saying that Abdoulaye proclaimed himself not Mahdī but more modestly 'master of his time' (*böröm jamano* in Wolof, *quṭb al-zaman* in Arabic),[29] and that he had not envisaged an immediate transfer of power but simply announced the future independence of Senegal and other African countries. Whatever the truth, the Mouride authorities managed to calm down the inspired *marabout* and succeeded in reassuring the colonial government. Abdoulaye lived out his days peacefully at Thiès, where he settled with his disciples in an area which still bears his name and where a number of *ṭālib*s live. The house he built still serves as a *zāwiya*.

Abdoulaye is thought to have had about twenty wives, but, curiously, he had only two male children; and both died in about 1939. The most eminent of his wives was doubtless Tabara Cisse (deceased in 1964), who was from a noble

[28] On the Bāy Falls of Cruise O'Brien, *The Mourides of Senegal*, pp. 141–58.
[29] Though his seal bears the title of Mahdī.

Mandinka family which became deeply involved in religious philosophy and, according to our sources, 'touched by grace'. The marriage had a peculiar significance in that it had been recommended by God to Abdoulaye in a revelation. Tabara Cisse was the mother of Sokhna Magat Diop.

When Abdoulaye died this daughter became responsible for her father's Mouride community. At the time she was married to Massamba M'backé (having been married, when very young, to Shaikh Ibra Fall). Her family explained her succession as being perfectly normal and 'legal' according to Mouride tradition: when the chief of a brotherhood has no male heir the oldest daughter takes over the title and function of *khalīfa.* Though the Mahdī of Thiès, lacking a male heir, might well have decided to appoint his best disciple, the ties of blood prevailed; among the Mourides, even more than in other brotherhoods, only a blood relation can inherit charisma. This succession was not, however, simply determined by Mouride tradition.

It appears that even before the *marabout*'s sons died he knew that his daughter Sokhna would assure the continuation of his mission. From birth Sokhna had been 'singled out' by God; and Abdoulaye Iyakhine gave her the religious education suitable for a religious leader. 'While my grandfather was alive', Mama M'Balla M'backé told us, 'he gave my mother lessons in conduct with the *ṭālib*s, the *marabouts*, and the colonial administration, in order that she might fully assume the succession after his death. He also imparted to her the secrets of God [*sirr* in Arabic, *mbot* in Wolof]'. In this way Abdoulaye prepared his daughter for the functions she still performs. He also transmitted to her his *baraka* and even his holiness. To her children and followers Sokhna Magat Diop is undoubtedly a *walī*, a 'friend of Allah'. According to Mama M'Balla, 'all sons and daughters of a *walī* inherit his *baraka*; but the holiness can only be communicated to the best of them'. However, as the 'blessing' is transmitted by the women too, Sokhna owes part of her *baraka* to her maternal family. She still has very special relations with them; and her younger son, Shaikh, who is a student at Jeddah, is tutored by Moustapha Cisse, Senegal's ambassador to Saudi Arabia.

Distinguished by this double seal of grace, Sokhna Magat

Diop has exceptional powers. Though even her son admits that these are neither as prodigious nor as spectacular as her father's, they are nevertheless so great that numerous *ṭālib*s come to ask for her intercession. Her 'secrets' and regular 'visions' are to her disciples proof both of her closeness to God (in other words a confirmation of her status as *walī*) and of her capacity to intervene with him on their behalf. However, her powers in this respect cannot be dissociated from those of her father. *Ṭālib*s who come to ask her a particular favour feel obliged, before seeing her, to retreat to Abdoulaye Iyakhine's shrine to pray for his help.

The *khalwa*s (mystic retreats) frequently practised by the *khalīfa* are supposed to help the disciples fulfil their wishes. The *marabout*'s *khalwa* may respond to precise and often very material demands, or it may be a strictly spiritual exercise in the adoration of and search for God. Sokhna Magat Diop is deemed to be the only woman in Senegal able to carry out this mystic practice, which is considered very dangerous (*seytan*s, in particular, subject anyone engaging in *khalwa* to terrible trials). It appears that the *marabout*'s *khalwa*s, which may last up to forty days (though more often three, five, seven, nine, or fifteen days), have become an important element in her holy status and reputation as an 'exceptional woman above men', of which we kept being told during our time in her *zāwiya*. Fewer and fewer *shaikh*s dare affront the perils inherent in this kind of exercise for the 'great initiates', which puts them 'face to face with God'. But the secret names of God known to Magat, as well as her ascetic life, favour her resort to this Sufi practice. Moreover, the *khalīfa* produces talismans and uses *sāfara* (water blessed through the dilution in it of Qur'anic verses written on paper, which is either consumed or used for ritual ablution) to help her *tālib*s (with healing, promotion, protection, and other needs).

This *marabout*'s authority over the brotherhood branch founded by her father has been consolidated by her many talents. Since 1943 she has presided over this Mouride community which, though not one of the largest of the *ṭarīqa*, may have several thousand disciples, not only at Thiès but also at Dakar, at M'Boro, and in the Banjul region of the Gambia. Her father's *ṭālib*s have confirmed their act of submission to her;

subsequently other disciples have given their allegiance to her. She has the normal powers of a *khalīfa* over her community. She can deliver the Mouride *wird* and appoint *muqaddams*; she receives the *adhiya* (tithe); she owns agricultural land (at Thiès and M'boro) which is worked by her followers; she 'possesses' several *dā'iras* (the largest being at Thiès); she counsels her disciples about important decisions in their lives (especially marriage); and despite her extreme discretion she is the 'strong man' of her area and the obvious intermediary of the Administration, for instance when local chiefs are appointed. Her disciples accord her all the respect, devotion, affection, and deference due to a *serigne* (*marabout*). Moreover, like all great *marabouts* she is active on behalf of her community and mobilizes her disciples to undertake collective works (they have constructed two mosques and, more importantly, her father's mausoleum). She has also instituted a *magal* in her father's memory, attended every year by most of her disciples, during which some forty marriages are celebrated.

Her status as *khalīfa* appears to be recognized by the Mouride hierarchy; and her father's errors now seem forgotten. Like other chiefs from great Mouride 'houses' Sokhana Magat Diop is under the direct authority of the Khalifa-General of the Mourides, Abdou Lahat M'backé, to whom she has declared her *jebellu* (as she had before to his predecessors). She accepts orders from him alone and executes these with remarkable zeal. To take an example, she raised funds for the completion of the great mosque at Touba; this earned her considerable praise from the Khalifa-General (she had collected some 2,000,000 CFA francs, equivalent to 40,000 French francs). Like most great Mouride *marabouts* she has a residence at Touba, the holy city, where she goes every year to attend the great Mouride *magal*. As is also the case within most great Mouride families, she has special marital links with the M'backé-M'backé family. Having made her second marriage with Massamba M'backé (her first husband having been Shaikh Ibra Fall), she made her third marriage, in 1945, with Shaikh Atta M'backé, the son of Amadu Bamba's older brother. Her last husband was himself *khalīfa* of the branch of his father Mamor Diarra M'backé and several of her children have married members of the M'backé-M'backé family. These matrimonial strategies provide additional evidence of the legitimacy of Serigne Abdoulaye's community and of that of its present chief within the brotherhood.

Everything seems to indicate that nothing distinguishes Sokhna Magat Diop from other great Mouride *marabouts*. However, a closer look reveals that her position as head of a brotherhood community gives it some special features. For instance there is her way of guiding her disciples. Being a woman, she cannot carry out certain tasks of a *khalīfa*. She never goes to the mosques which she helps construct and she never conducts public prayers. On the contrary, she retreats every morning to her father's mausoleum. Nor does she carry out legal functions, such as the celebration of baptism or marriage. However, these limitations do not prevent her from intervening in matters related to the mosque or Islamic law. She appoints the *imāms* and she arranges the marriages of her *ṭālibs* even if it is an *imām* or his eldest son who officiates.

Sokhna never performs in public. During the *magal* that commemorates her father it is her eldest son and future heir Mama M'Balla M'backé who makes the customary speech and replies to the invited members of the Administration. But he still conceives himself as a mere spokesman for his mother and follows her guidelines scrupulously. On most public occasions the eldest son clearly plays an important role; but he considers himself merely the *loxu-serin-bi*, or arm of the *marabout*, even though he already has his own *ṭālibs*. One should not be led to believe that this relative public invisibility of the *khalīfa* indicates some sort of limit to her authority. On the contrary, her discretion is interpreted as a sign of her holiness. It is not only because she is a woman that Sokhna shows such reticence, but especially because she is a *walī* whose *baraka* transcends 'ordinary' manifestations of authority.

Sokhna lives as a virtual recluse in her separate apartments; she rarely goes out, travels little, and is seen in public only at very special occasions. She spends most of her time praying, reading Amadu Bamba's *qaṣā'id*, and practising *khalwa*. She eats little and talks in a low voice; her visitors are strictly screened. She has adopted the kind of mystic lifestyle that is becoming less and less common among Mouride *marabouts*. This is what makes her so outstanding within her community and distinguishes her most clearly from other Mouride *marabouts* who tend to become increasingly 'worldly'. Hence it is less important that Sokhna lacks the legal powers of a man. This only makes her more holy; the fact that she does not pray in the mosque but at home makes people in the *zāwiya* regard her prayers as superior religious acts. We were told that 'to adore God, it is best to be alone'; and someone recalled the Prophet's saying: 'Praying alone is twenty-seven times better than praying at the mosque.'

It seems that the presence of a woman at the head of this Mouride branch has accentuated the Sufi dimension of the holy authority. Could it be that mysticism, in erasing or modifying the difference between the sexes, points the way towards the acceptance of female authority in Islam? Despite her determined claim to be her father's heir, Sokhna Magat Diop is far removed from the Mahdist inclinations of Serigne Abdoulaye and from the eccentricities of the Bāy Fall; though this is where the origins of her community are to be found. Under her authority the Bāy Fall culture has been superseded by a more typically Sufi style. This evolution is distinctly from that of the average Mouride brotherhood, where mysticism is increasingly effaced by considerations of business and power.

The *khalīfa* apart, her community is also generally inclined to value women and promote their place in Islam. A whole section of the *zāwiya* is concerned with justifying sexual equality in religious life, stressing that 'in Islam, women have the same responsibilities as men', and that 'there is no distinction between men and women in the worship of God'. It is hardly fortuitous that several women hold key positions in this community, whether as *muqaddam*s or in the *dā'ira*s. The main *dā'ira* which unites women and men (though the latter are fewer in number) is actually directed by a daughter of Sokhna Magat Diop, N'Deye, even though for the sake of form she has a male co-president. Another *dā'ira* has been named after Amadu Bamba's mother Mame Diarra Bousso, and takes an active part in the annual pilgrimage in her memory to Porokhane. Sokhna's community displays particular reverence towards this important Mouride figure, and has been promoting the cult commemorating her in recent years.

Contrary to what a first impression indicates, the type of authority Sokhna exercises and the religious life of her community provide clear evidence of an Islamic practice and style which have evolved because the *khalīfa* is a woman. Over the years this Mouride branch has evolved in a unique direction which distinguishes it from other Muslim groups. Although it may be exaggerated to talk of a feminist Muslim community, you need only spend a few days in Sokhna's *zāwiya* to become aware of a more feminine flavour than one generally encounters in the Mouride brotherhood.

## CONCLUSION

It would be wrong to over-emphasize the singularity of this 'female' Mouride branch. In the final analysis its peculiarities, albeit institutionalized or at least 'imposed' in a rather artificial way, exist also in other Islamic groups in more or less attenuated forms (as noted in part 1 of this study). The Sufi tradition does not insist upon sexual separation in religious expression; the cult of saints and *marabouts* does not inherently exclude women; and female initiative is a black African tradition. Hence African women are never entirely relegated to a peripheral or semi-pagan religious space.

Moreover, the tradition of female associations and female financial independence makes black African women desirable partners in any institution (be it religious or not) which aims at mobilizing people and wealth. The *marabouts* of Senegal show their awareness of this when they encourage female *dā'ira*s, allow women to be independent *ṭālib*s, and authorize a woman *khalīfa*. These recent developments show that women may increasingly provide the key to the future of the brotherhoods: they are remarkable religious agents. However, one wonders whether, as bearers or as organizers of *baraka*, they will be able to remake the *ṭarīqa*.

# 7

# Charisma Comes to Town
# Mouride Urbanization 1945–1986

DONAL B. CRUISE O'BRIEN

AGRARIAN crisis in northern Senegal has prompted beleaguered Mouride disciples to seek economic salvation in the town. The urban drift, even where it is on a seasonal rather than a permanent basis, remains a minority phenomenon for the brotherhood as a whole; but recent national census statistics show the highest rates of rural–urban migration to be from the Mouride regions. Drought in recent times is part of the explanation: the rains have often failed over the past twenty years, and the visual evidence of the desert's southward march is clearly to be seen in northern Kayor and Baol.[1] Government policy in managing the agricultural sector, with its overtaxing of the cash-crop producer, has also contributed to the discouragement of those who would earn a living from the land. But, whatever may be the reasons for the population movement away from the country and into the city, this movement foreshadows a significant shift in the organization of the Mouride brotherhood. Threat and promise are both included in the likely future possibilities for the brotherhood.

Urbanization would seem to be an obvious threat to the rural saintly establishment, at least in the medium term. The saints already find their country estates to be of declining profitability, but they have chosen not to follow the more enterprising or

My thanks are due to the British Academy and to the Nuffield Foundation for their financial support of research in Senegal (1980–1, 1983), and to the School of Oriental and African Studies for granting research leave. The School also granted leave for research among the Parisian Mourides (March–April 1985). Particular help came from Thierno Sow and Gerard Salem in Dakar, and from Mustapha Diop and Daniel Ferrat in Paris.

[1] The administration regions of Louga and Diourbel correspond approximately to the traditional Wolof units of northern Kayor (Ndiambour) and Baol. Recent statistics on migration out of this area were mentioned by John Waterbury in correspondence from Dakar and Princeton; see M. Gersovitz and J. Waterbury (eds.), *The Political Economy of Risk and Choice in Senegal* London: Frank Cass, 1987).

desperate of the disciples in moving to Dakar. For the moment the Mouride *shaikh*s seem to be settling for a comfortable parasitic existence around the brotherhood's village capital, Touba. So the new city-dwelling disciples, abandoned by their traditional leaders, have been forced to develop their own urban organization. The urban Mourides have responded energetically to the challenge of their new situation: they have developed a viable associative organization, acting under the umbrella authority of the brotherhood's Khalifa-General, but with a new urban leadership.

The country saints now find their authority to be threatened by the rival leadership emerging in the town. Worse, the saints are challenged from two sides in the town—from one side by the brotherhood's more successful traders, hard-faced men who can be very much richer than the *shaikh*s, and from the other by the new generation of Western-educated urban disciples. The growth of a Mouride intelligentsia may be an even more dangerous development for saintly authority; for the intellectuals see themselves as the best heirs to the charismatic legacy of the founder, Amadu Bamba. They can interpret the founder's written word; they can be the most effective propagandists on the founder's behalf; they can threaten at least implicitly to render the traditional *shaikh* superfluous. The Mouride intellectuals' skill in sacred agitation and propaganda could then even bring about a whole new phase of the brotherhood's expansion—both within Senegal and beyond the national frontiers. Such has been the promise.

The recent development of an urban Mouride community may perhaps best be understood in terms of two models, one of introversion and the other of extroversion. The organization of Mouride traders, sociologically inward-looking, reflects the understanding that the brotherhood's commercial community thrives best with the most limited social contacts outside of the brotherhood's own networks. Mouride ritual in this case serves as the bond of a commercial association; weekly sessions of communal singing allow the businessmen singers to remember the founder through his chanted words. No attempt here is made to win new converts to the brotherhood; on the contrary the priority is to exclude the non-Mouride outsider—such is the social function of the ritual. The singers may not precisely understand the words they so noisily sing, but they praise the miracle-working Amadu Bamba as the bearer of salvation in the next world while they use his holy name to establish their boundaries of commercial trust here below.

The other (extroverted) model of organization is that of the modern Mouride intelligentsia. Here proselytism is the central purpose, to make the founder 'better known' through the written and the spoken word, through the magazine and the public meeting (not to speak of the film and the cassette). Amadu Bamba is here remembered as the author of sacred texts which can be translated, understood, and made the subject of textual commentary. He becomes less the worker of miracles (in the words of the Dakar Mouride student leader, '*Il faut dépasser le miracle*'[2]) than the exemplar of a particular stance in the face of Westernization. We witness here the formation of a Mouride ideology. Amadu Bamba is still the saviour of his disciples, the guarantor of paradise, but he is also presented as 'the saviour of Islam' in Senegal, as representative of 'the dignity of the black race', and as pre-eminent 'hero of the Senegalese nation'.[3] This ideological version of Amadu Bamba is potentially subversive of the brotherhood's traditional leadership. The intellectuals are putting forward their own candidacy for the brotherhood's direction, as masters of the written word. They can claim to know Amadu Bamba best, through his translated writings, while they can also claim to be best qualified to make the founder known, through their modern organizational capabilities. These intellectuals are also well placed to defend the Mouride brotherhood against its modern critics, inside or outside of Senegal.

Amadu Bamba's charisma, while thus predictably in course of routinization, is by no means exhausted; an urban pilgrimage (annual since 1980) and an urban miracle (1984) attest to his enduring extraordinary powers. The Mouride Khalifa-General, as the founder's modern heir, must now work to see that the challenges of urbanization can be worked to the brotherhood's advantage.

[2] Interview with Saliou Mboup, President of the Dā'ira des Étudiants et Élèves Mourides du Sénégal (DEEMS), Dakar (Jan. 1981).

[3] Phrases quoted from the Mouride intellectuals' paper *Ndigel: La Voix du Mouride*, over the period 1981–4.

## 1. THE SECULAR PURPOSES OF SACRED RITUAL: THE COMMERCIAL MOURIDES

The range of Mouride commercial migration has extended over time, and a rough division into historical phases should perhaps be made. The first urban Mouride settlement was organized by Shaikh Ibra Fall in the territorial capital Saint-Louis, a centre of Mouride diplomacy as well as of petty trade at the time of the brotherhood's earliest development (from 1892). It was here that Ibra Fall acted on the founder's behalf, with a flamboyant if unconvincing humility, in his difficult relations with the colonial authorities. Ibra Fall also used the influence of Senegal's parliamentary deputy to effect Amadu Bamba's return from Gabonese exile (1902). At this time the economic energies of the great majority of Mouride disciples were of course devoted to agricultural settlement, to millet and ground-nut cultivation, and it was not until the 1930s that some commercially minded disciples began to set up shop by the railway halts along the Dakar–Niger line; L. Nekkach dates this movement from 1934 to 1945.[4] One of the founder's brothers, Shaikh Anta M'backé, incarnated the commercial spirit: 'He was a businessman, he wasn't like Amadu Bamba. He was the sort of man who comes and goes, who is always travelling. He had houses everywhere and connections with everyone.'[5] He also bought three complete bedroom suites from the entrepreneur and politician Galandou Diouf and rescued the latter from bankruptcy.

A second phase of the disciple's urban migration and settlement opened in 1945, directed principally towards the city of Dakar. Dakar became the focus of Mouride enterprise, especially after national independence (1960). The 1960s were years of consolidation of the Mouride community in the capital. The newly urban disciples of this time were still distrusted by their country cousins, being suspected of disloyalty to the saints and to the brotherhood.[6] Only with time has it become clear

[4] L. Nekkach, 'La Mouridisme depuis 1912', unpublished paper in *Archives de la République du Sénégal*, Dakar (Saint-Louis du Sénégal, 1952), 8.

[5] Shaikh Yaba Diop, quoted in J. Roch, 'Entretiens avec des marabouts et des paysans du Baol', roneo (Dakar: ORSTOM, 1968).

[6] The mood of the time is reflected in the chapter 'The Urban Brothers' in D. B. Cruise O'Brien, *The Mourides of Senegal: The Political and Economic Organization of an Islamic Brotherhood* (Oxford: Clarendon Press, 1971), 237–61.

both that the urban Mourides have successfully maintained their Mouride identity, and indeed that they have turned that identity into a formidable commercial asset. Thus the principal open-air market of Dakar (Sandaga) has progressively been 'conquered' by Mouride tradesmen, a substantial achievement when one recalls that these traders have only been arriving since 1960, coming with scant experience of commerce or of the city.

A third phase of Mouride urbanization may be discerned from 1966 onwards, as some of the brothers began to explore the possibilities of street-peddling in France, even pioneering their trade in the other states of the European Economic Community. The Paris *Exposition de l'Art Nègre* in 1966 appears to have inspired the Mouride salesmen to make a speciality of 'African art objects'—masks, beads, bags, and carvings of all sorts.[7] It should be noted that the dating of these 'phases' in Mouride urbanization, while possibly useful, is also somewhat arbitrary. The dividing lines are rather more blurred in reality, notably by the shift from temporary (seasonal) migration to permanent urban settlement: you go to town during the dry season, peddling what you can; but if the rains don't come at sowing-time, or come in insufficient quantity, you stay where you are—in town.

The formal organization of the brotherhood's urban life is deceptively simple, hinged on the association or *dā'ira*. The formal purpose of such associations is to organize weekly sessions of religious singing, and to collect the members' subscriptions. These fees are then devoted in part to the purposes of mutual assistance, in part to the expenses of the brotherhood as a whole, through the Khalifa-General as overall leader. Tentatively questioned on the possible help which such a pious association might provide in matters of commerce or of relations with state officialdom, *dā'ira* members respond with protestations of the associations's innocence. One may have some difficulty in accepting such disclaimers. I well remember a Pikine *dā'ira* in 1981 where the association's officials told of their own innocence

[7] See the valuable work of Gerard Salem, now of ORSTOM, Dakar: 'De Dakar à Paris: Des diasporas d'artisans et de commerçants', *Doctorat de troisième cycle* (Université de Paris, 1981); id., 'De la brousse sénégalaise au Boul'Mich: Le Système commercial mouride en France', *Cahiers d'études africaines*, 21/1–3 (1983).

of the things of this world, while another party waited in the yard—the *Prefet* of the *département*.

One may err, none the less, in too sceptical an assessment of the Mouride association's other-worldly purposes. The *dā'ira* truly is devoted above all to those weekly songs, and as an association it does not in truth need to do more in order to accomplish something of substantial worldly value for its members. This ritual activity helps to draw the boundaries of an urban Mouride community, boundaries which include the Mouride traders and which also exclude the commercial competition. The *dā'ira* gives organization and structure to a moral community with the insistently recalled memory of the brotherhood's founder. Members of that community then begin to enjoy a comparative advantage in the competitive strife of shanty-town commerce—the advantage of reciprocal trust. The point here is that Mourides in business can trust each other that little bit more, as their network of social contacts lies more exclusively within the brotherhood. Outsiders have learned to respect Mouride commercial competition: as they put it, using the vocabulary of football, 'in business, you mustn't let yourself be dribbled by the Mourides'.[8] Which is not to imply an unqualified solidarity among Mouride traders; but if a Mouride trader has the choice of cheating another Mouride or a stranger, he'll cheat the stranger. We may thus use Abner Cohen's blueprint in identifying the urban Mourides as a trading diaspora.[9]

The brotherhood's associations may be understood as providing an urban network where the brothers learn the arts and the advantages of associating together. Here the newcomer to the city can be welcomed and assisted in many practical ways—to find lodgings, to receive basic commercial advice, perhaps even a little credit. He is encouraged to see himself as part of a united social front which breaks down the status hierarchy of traditional Wolof society. No more nonsense now about lowly social origins, casted or slave; 'We're just Mourides

[8] Salem, 'De la brousse sénégalaise au Boul'Mich', p. 278.

[9] See A. Cohen, *Custom and Politics in Urban Africa: A Study of Hausa Migrants in Yoruba Towns* (London: Routledge and Kegan Paul, 1969); id., 'Cultural Strategies in the Organisation of Trading Diasporas', in C. Meillassoux (ed.), *The Development of Indigenous Trade and Markets in West Africa* (Oxford: Oxford University Press and International African Institute, 1971).

now.' And to be 'just a Mouride' is to be a person who counts for something in the urban world, a person respected and sometimes even feared, someone with lots of confidence both in his social standing and in his commercial efficiency. Such confidence is of course a mighty asset in the commercial world.

The urban association, while it does structure its members' involvement in a total Mouride community, makes no attempt to channel the individual enterprise of its members. The association does not in general lend money to its members, for fear of default. But individual friendship groups within the association can of course lead to commercial undertakings, and when one member lends money to another the penalty for default may be understood to have some of the force of the sacred. There is also the insurance policy implicit in brotherhood membership, the promise of aid in the event of personal misfortune, and of course the promise of paradise beyond.

The commercial patronage of eminent Mouride businessmen may also be sought by the ambitious trader. The successful Mouride entrepreneur appears to have taken on something of the social function of the rural *shaikh* in providing for the material existence of his clientele. Clientelism in the Mouride style operates, of course, on stiff terms: in the city it can mean years of street-selling, on the tightest of commercial margins, on behalf of a Mouride wholesale dealer. But that wholesale dealer offers various forms of valued assistance to his retailers, whether in procuring official papers, in finding lodgings, or in funding a trip. And the retailers talk of their business patrons very much in the language of total obeisance which Mourides used previously to reserve for the brotherhood's holy men, although the businessmen, however successful, have no religious authority.

Mouride traders seem to provide us with what is in its own terms a commercial success story, although it is a difficult tale to document or to buttress with figures. Senegalese official statistics since independence have carefully avoided any breakdown by brotherhood affiliation for fear of communalist political reactions, while the brothers themselves offer 'statistics' only in the largest and most evidently propagandist of approximations. With a good deal of Mouride commerce being conducted around the margins of the law, across international frontiers, and around administrative regulations, detailed questioning by an outsider is scarcely tolerated. For an overall assessment there does remain plenty of visual evidence, for example in the striking proliferation of vans and buses in Dakar bearing the word TOUBA; and there remain also the verdicts of the best informed outsiders. Evidence of this kind indicates that the Mouride traders in Dakar have been steadily growing in strength over the past twenty-five years; they may now control more than half of the city's retail trade.[10]

Mouride traders in France, bearers of trinkets, leather goods, and 'African art objects', have defeated all African competition over the past decade. The market in African art being by now more than adequately supplied, the brothers will need to diversify. Some of the younger traders in Paris appear to have become involved at least temporarily in the more lucrative but dangerous market in drugs. (There were several rounds of arrests in 1983–4 at the market near the Gare de Lyon, a favourite of the brotherhood's traders, although the brotherhood's own investigation established that 'those who were arrested were not Mourides'.[11]) Others, in Marseilles, have specialized in cloth. The brotherhood's traders must generally operate in a hostile environment, whether in France or elsewhere in Europe. This is a lonely, unstable life of poverty, marked by racial incidents and difficulties with the police. These hardships are indeed, as Gerard Salem has remarked, comparable in their way to the difficulties endured in past times by the agrarian settlers of the *dara* pioneer farms.[12] The traders are now the brotherhood's pioneers; their quest for new markets now embraces all the main French cities, and they are spreading to Italy, to the Low Countries, and to Germany on a more or less experimental basis. How have they done it? By their energy and their organization, clearly, and their willingness to endure hardship; but then we must refer back to the fervent religious belief which has made the brotherhood so effective a social force. It is this religious fervour, the shared belief in one man's charismatic gift, which is at the bottom of this commercial success story.

[10] Interview with Matar Diop, Secretary General of the governing Parti Socialiste, Dakar (Apr. 1983).

[11] Interview with Mustapha Diop, Paris (1985).

[12] Salem, 'De Dakar à Paris', p. 132.

Political power also helps, of course; the brotherhood's power in Senegal can be converted into a commercial asset. A relative ease of access to state credit and a relative immunity from official administrative harassment accompany national political power. The Khalifa-General is the brotherhood's principal power-broker, and he can provide very useful patronage for commercial enterprise, as in the case of the free-trade community which has long flourished in Touba.[13] For their part, the Khalifa-General and some of the leading saints derive a substantial cash income from the businessmen's devotional tributes.

Yet the alliance of interests between *marabouts* and businessmen may not be without its tensions; the abortive outcome of the government's 'clean up Touba' campaign of 1977 has been interpreted as a victory of the Mouride smuggler-traders over the *marabouts*. And with the attempted murder of the *imam* of the great mosque (Serigne Abdou Qadir M'backé, the Khalifa-General's younger brother and Amadu Bamba's son) the life of crime appears to have caught up with Touba. Firearms have become increasingly commonplace in the holy city over the past decade, and this last attempted murder (although not involving firearms) has provided the Government with an unanswerable pretext for the installation of a brigade of thirty gendarmes in Touba, the first substantial police presence there since Senegalese independence. It is probably too early to see this presence as '*la fin de la fraude*';[14] we may well be in a shadow-zone of disguised intent. In any case it is clear that despite the rise of crime in their domain the successful Mouride businessmen have already become a powerful counter-élite within the brotherhood.

Lastly, we must take note of the urban miracle: in 1984 a fashionable lady, examining her dress in the mirror of a Dakar boutique, saw not her own reflection but that of Amadu Bamba.

[13] Since 1985 Touba market has been patrolled by a few policemen, who have even made some arrests. This surveillance may well be related to the intriguing question of the 'disappearance' of almost 300,000 tons of peanuts from the national harvest of 1984/5; see the national newspaper *Le Soleil*, 11 Mar. 1985, 'Où sont passées les graines?'. The information on Touba's police is from an interview with Babacar Lo of the Mouvement Islamique des Mourides d'Europe, London (May 1985).

[14] *Africa international* (Dakar), 185 (July–Aug. 1986). See also Momar Coumba Diop, 'La Confrérie mouride: Organisation politique et mode d'implantation urbaine', *Doctorat de troisième cycle* (Université de Lyon II, 1980); id., 'Les Affaires mourides à Dakar', *Politique africaine*, 1/4 (Nov. 1981).

The ensuing hullabaloo naturally brought the local Mouride organizers to the sacred scene, where they promptly insisted on buying the shop. This is after all the first Mouride miracle to have been recorded in Senegal's capital city; the shop can be expected to become a place of pilgrimage, a notable resource in the power-play of Senegalese Sufism. Interpreted in secular terms, such a miracle represents the Mouride assertion of a permanent urban presence: the brothers want it to be known that they have come to town, for keeps.

## MOURIDISM AS IDEOLOGY: THE EMERGENCE OF AN INTELLIGENTSIA

'French' education was held to be the path to damnation in the time of the last Khalifa-General (Falilou M'backé, who held the title from 1945 to 1968). Disciples were encouraged to shun the French (language) school and the French university. Did not Amadu Bamba's sufferings at French hands, his 'thirty-three years of exile', dictate a reticence towards France? Few Mouride children were entered for the *lycée*, very few indeed for the university, and the exceptions were most discreet. This situation has, however, changed under the present Khalifa-General, Abdou Lahat M'backé, as the disadvantages of such a self-imposed exclusion from modern higher education have been recognized: why should Mourides have to accept exclusion from senior levels of employment, either in the professions, the large corporations, or the civil service? If education has been the barrier, then it is a barrier to be overcome. The State is encouraged by the Khalifa-General to open schools in Mouride areas, and Mouride parents are encouraged to enrol their children. A conspicuous outcome of this shift has been the emergence of an assertive Mouride presence in the university, in the form of militant Mouride students with an active association, a programme of proselytism among the young, and an ideological restatement of Mouride faith. Amadu Bamba here becomes not only a source of textual authority for the disciples: he is an example to Islam and to the Senegalese nation, as pioneer of the struggle against Western decadence.

Mouride student enthusiasm and the mobilization of the

younger generation appeared at one time to be heading in dangerous directions. Aggressive Mouride proselytism among the urban young, spreading outwards from the university and the *lycée*, heightened popular awareness of differences between Senegal's Sufi brotherhoods; thus an explosive communalist politics was prepared in the late 1970s. The students' apparently innocuous purpose may have been merely to make Amadu Bamba 'better known', with a series of highly popular public lectures on the theme of 'Mouridism and youth'. But it was Mouride student self-righteousness, with a more or less explicit critique of other brotherhoods, which awakened the spirit of communalism. As the director of the Senegalese Tijāniyya remarked in 1979: 'There are things happening in this country which, unless they are corrected, could lead to civil war.'[15] Abdul Aziz Sy followed this warning with an appeal 'to the learned and to the leaders of brotherhoods' to guide the Islamic community away from factionalism. The Mouride Khalifa-General, not a particularly learned man for a brotherhood leader, duly responded in 1980 with a call for unity in the *umma*. The Khalifa's appeal does seem to have been remarkably effective in restraining the over-zealous disciples.

The origin of this dangerous student mood should apparently be sought in the *lycées* of Saint-Louis, in the year 1975/6, when a small group of pupils met to sing Amadu Bamba's devotional poetry and to drink tea. They met under the tuition of a Mouride *marabout*, Abdoulaye Sarr, and there were one or two conversions to Mouridism (from the Tijāniyya). The idea of actively Mouride *lycéens* was at this time strange enough for them to be pointed out in the street as objects of general surprise. Upon leaving school some of this group enrolled as students in Dakar University, and in 1977 they asked the Khalifa-General for authority to form a Mouride student association—a *dā'ira*. The Khalifa apparently made this very much his personal concern, instructing the students in the basic Islamic practices as well as in Mouride mysticism, using the founder's poetry as his text.[16]

This student association of course prospered mightily, linked soon after with a national federation of young Mouride

[15] Speech in Tivouane (1979), quoted in Diop, 'La Confrérie mouride', p. 191.
[16] Interviews with Saliou Mboup and Mamadou Dièye, Dakar (Jan. 1981).

associations. Young Mourides vaunted their own superior organization as conversions multiplied, leading to a mood of giddy triumphalism—'everyone will end up Mouride', said the intellectuals' magazine.[17] In part this may be seen as the anti-Western reaction of the times, a Mouride puritanism (against pop music, the use of drugs, Western dress, in short against the 'state of debauchery of our youth'). Amadu Bamba was held up as the example and the instructor for the young, and the national youth association organized an 'Amadu Bamba cultural week' between 24 and 31 December 1979, partly in protest against the 'seasonal debauch' of Christmas. This was a cultural week with a political echo in a Muslim country which at that time had a Roman Catholic president, Leopold Senghor.

The saintly Mouride founder, redeemer of his disciples and author of numerous gaudy miracles, has been transformed in the university representation of his achievement. The miracles of rural popular belief (such as escape from a hungry lion or a raging bull and prayer on the ocean waves) have become a matter for embarrassment—'*ces miracles sont à la portée de n'importe qui*' (those miracles are within anyone's reach).[18] The Mouride intelligentsia prefers to emphasize the founder's literary and pedagogic achievement, and they insistently repeat Amadu Bamba's declaration: 'my real miracles are my writings'. The founder's Arabic poetry is translated, into Wolof and French, and used for religious instruction. Amadu Bamba remains the symbolic centre, remembered as 'the man who said no to the Whites'. School associations apparently love to hear the story of how Amadu Bamba refused to deny his Muslim faith before the Governor of Senegal, whereupon he was exiled by the colonialists (in 1895), with the delicious implication that those other Muslim leaders (Tijānis and Qādiris) who were not exiled must have denied their faith. Ideas like that could indeed have led to civil war.

The arrival of a Mouride missionary in the university world of Paris was also to be the occasion for a remarkable flurry of Mouride proselytism. Abdoulaye Dièye arrived at the *Maison des Étudiants Africains* in 1977, a mature student in architecture and landscaping who was to combine his studies with the most active

[17] *Ndigel*, 2 (n.d.).
[18] *Ndigel*, 9 (CEE Bruxelles, 3–4 July 1982).

work of Mouride propaganda. Dièye came from a St Louis Mouride family, and he had enough personal confidence to restate the founder's message in a discouraging environment. The African students' house had for some time been the scene of leftist opposition politics, with a scornful rejection of the opium of the people. When Dièye started to spread his Mouride message he was thus darkly suspected: was he an agent of Senghor, or of the comprador bourgeoisie, or was he just an obscurantist?[19] He had no time for leftist disputes ('intellectual masturbation') or for opposition politics. He proposed Mouride Islam as an alternative to 'cultural alienation': first the basic principles of Islam, then the higher mysticism. He 'gave of himself without stint. He sacrificed his family to teach us Arabic, the Qur'ān, and the thought of Amadu Bamba . . . there were always people in his rooms'.[20]

Dièye's assessment of the students suggested an approach through archival investigation; as he put it, 'documents were their Qur'ān'. They were encouraged to search the archives of the colonial administration, and then to study the many theses and published works on Mouridism, so that they might 'restore the truth' about Amadu Bamba. The truth to be uncovered was that he was 'a good nationalist, a leader of men',[21] who had been singled out by the colonialists as their principal adversary—hence the exiles and the persecutions. Armed with such ideas, the Paris students were encouraged to 'live the thought' of Amadu Bamba. Then they, rather than the 'Senegalese gangs' who had 'perverted' the founder's thought, could present themselves as the true intercessors. They could express the founder's thought in its authentic Muslim orthodoxy and universality. So the Paris students established their own Mouride association in 1977, the Association des Étudiants et Stagiaires Mourides d'Europe (AESME), which soon established branches in the principal French university centres. This was allowed by the creation of a Mouride magazine, *Ndigel: La Voix du Mouride*, which claimed a circulation of one thousand

[19] See Mustapha Diop, 'Les Associations murid et la da'wa en France', unpublished paper delivered in Feb. 1985 to conference on Muslim minorities in France, Institut d'Études Politiques, Paris, p. 5. This paper provides an excellent introduction to the Paris Mourides.

[20] Ibid., p. 6.

[21] *Ndigel*, 12 (Nov.–Dec. 1983).

copies by its fourth number. The magazine was trilingual—printed in French, Wolof in Arabic script, and Arabic. The contents reflected the outlook of the editor and principal contributor, Abdoulaye Dièye: much basic Islamic education, always linked with the person and the work of Amadu Bamba ('Amadu Bamba and the *sharī'a*').

The explicitly political was not to be found in the pages of *Ndigel*, with the exception of a recurrent, soothing theme, 'Amadu Bamba and non-violence' (although the presentation of Mouride history has its political implications, to which we shall return). The purpose of this magazine was first to overcome student reserve towards the 'reactionary' Mouride brotherhood, and then to gain the support of members of other Sufi brotherhoods. Through the pages of *Ndigel* Mourides could address the hostile outside world—the Tijānis in Senegal, the Arabs in Europe, and the French in France—on the themes of Amadu Bamba's Islamic orthodoxy and his universal appeal. The Khalifa-General had of course given his prior approval both to the AESME and to *Ndigel*, being doubtless well aware that these initiatives held a major promise of the brotherhood's further expansion.

Such a promise of expansion appeared to be justified by Mouride appearances at international gatherings, first at UNESCO headquarters in Paris for the Journée Culturelle Amadou Bamba (1979 and 1981), then at the international Islamic conference in Kuala Lumpur to celebrate the fifteenth century of the *hijra* (December 1981), to which Abdoulaye Dièye was a delegate, and then at the European Economic Community headquarters in Brussels for the Weekend Culturel Amadou Bamba (3–4 July 1982). This last gathering was the occasion of much talk that might readily be forgotten, but also of a highly memorable address by the President of AESME, in which Serigne Sarr Diop announced that his association represented 'Islam which escapes the racial frontiers'. 'Although we are neither Arabs nor Arabic-speakers', he declared, 'we will act as a transmission belt between Shaikh Amadu Bamba and Europe, making the same approach as those who were, alas, unable to get beyond Poitiers in the year 732.[22] Fighting talk in

[22] *Ndigel*, 9 (CEE Bruxelles, 3–4 July 1982).

the capital of Europe, for an African delegate to evoke nostalgia for the attempted conquest of France by medieval Islamic *jihād*. Can this be the result of asking Mouride students to study historical documents?

The offended pride of black African Muslims can be seen in the insistent call to be recognized as full members of the *umma*. In the words of Abdoulaye Dièye (from a recorded interview):

> Arab Muslims come and play games with us, saying that Muhammad was an Arab. And what would be our lot if the founder of Mouridism had been an Arab? They would see us as slaves, because there are some Arabs, although they are Muslims, who scorn the Africans even more sordidly than do the French. We have two battles to fight, an external battle with the Westerners and an internal battle within the *umma* saying to the Arabs—you are the worst Muslims on earth, because if anyone has not understood Muhammad's message it is the Arabs.[23]

And on another occasion (in *Ndigel*):

> Is it not scandalous today to see those who claim to serve the best of prophets being led by perverts or atheists? Is it not absurd that those who call themselves brothers of God are turning their lands into powder kegs to destroy the *umma*?[24]

The prescribed corrective to such aberrations is to be found, of course, in studying the life and work of Amadu Bamba, Servant of the Prophet.

In the ideological formulation of the modern Mouride intellectual Amadu Bamba is presented as a figure of exemplary pride, 'l'Homme du Refus'. Refusal firstly of the politics of pre-colonial Wolof society, as when he declined the post of *qāḍī* in Kayor (1882), rejecting the world of the semi-pagan war-bands. Refusal secondly (and principally) before the arrogance of European imperialism, as when he was arrested in 1895, and said his Muslim prayers in the colonial governor's palace before being exiled to Gabon. Refusal thirdly before the racial scorn which he encountered while on his second exile in Mauritania (1903–7); among the Moors he demonstrated his intellectual prowess as well as his Muslim piety, thus confounding the French colonialists who had 'hoped to make him lose all his

[23] Quoted in D. Ferrat, 'La Confrérie des Mourides en France', unpublished *mémoire* (Institut d'Études Politiques d'Aix-en-Provence, 1985), 82–3.

[24] *Ndigel* (n.d.).

prestige before the learned Mauritanians'[25] (of the Sidiya family, his hosts in exile).

This powerful idea, of Amadu Bamba as 'the Man of Refusal', underlies the organization of a new Mouride pilgrimage (annual since 1980). The *Magal* of the two *Rak'as*, held in September in Saint-Louis, celebrates the occasion (5 September 1895) when Amadu Bamba was brought to the colonial governor's palace before being sent into exile, when he insisted on saying his prayers in the governor's office before calmly asking why he had been summoned. Then, in the words of the pilgrimage organizer:

> The colonial authorities understood that the black race was not as it had been described by their philosophers, their explorers, their *Commandants de Cercle*, and their governors. Before them was a man who was intelligent, proud, sure of himself . . . The objectives of colonialism were clear. The Senegalese must be oriented towards the Christian religion. The black man should be humiliated, and the Muslim leaders intimidated.

And in these testing circumstances,

> He alone dared to pray. He invited the others to remember, to witness that God was the only Master, that Muhammad was the Prophet of God.[26]

His prayers, the two *Rak'as*, were furthermore pointedly in the shortened form prescribed for Muslims in enemy territory—an additional note of defiance which may have escaped the governor and his associates.

According to French historical records Amadu Bamba was sent captive into exile after his summons to Saint-Louis. The founder's own mystic account is that he left voluntarily, in the company of the Prophet's companions (the immortals of the battle of Badr):

> I left with the pious (the immortals) when I had to leave for exile. The enemies thought that I was going as a prisoner. Thanks to the immortals of Badr I gained a military victory, and separation from the enemies of God.[27]

[25] *Ndigel*, UNESCO special number (4 July 1981).
[26] Quoted in *Ndigel*, 13 (Mar.–Apr. 1984).
[27] *Ndigel*, 11 (June–July 1983), and 13 (Mar.–Apr. 1984).

Today's Mourides believe that the founder's defiant prayers in St Louis were his protection before the various ordeals inflicted by his colonialist gaolers during his 'seven years, seven months, and seven days in Gabon'.

The *Magal* of the two *Rak'as* is the first pilgrimage to be organized by the Mouride intellectuals. The Khalifa-General is 'Honorary President' of the organizing committee and the founder's youngest son (Mourtada M'backé) is 'Acting President', but the real organizer was a local schoolteacher (Madike Wade). This *magal* appears to have been initiated in 1980, after a delegation of young Saint-Louisian disciples was received in Touba by the Khalifa-General. The young *ṭālib*s were accompanied on this occasion (27 July 1980) by some elder Mouride dignitaries of the town, and the Khalifa was sufficiently reassured to give his authorization to the project.[28] The pilgrimage lasts for a week, around the central day of 5 September (the date of the founder's convocation in 1895). The use of a 'Christian' date for a Mouride pilgrimage is in itself an innovation. Also new is the educational dimension of this pilgrimage—an exhibition of Amadu Bamba's work and a week of lectures on pious themes (e.g. 'Amadu Bamba and the Unity of Islam', 'Amadu Bamba and the Qur'ān').[29] And the whole pilgrimage concludes, rather splendidly, with a march on the Governor's palace at the Place Faidherbe (with much congestion in the office where the famous prayer is said to have taken place).[30] The pilgrims are drawn from the hinterland of Saint-Louis (the Région du Fleuve), and the occasion has become important enough to attract several ambassadors of African states and representatives of Senegalese opposition political parties, as well as representatives of Qādiriyya and the Tijāniyya. But above all this is an occasion for the young Mourides, celebrating not only a holy man, but also a racial and a national hero (*'notre héros national'*, in the words of Madike Wade at the 1984 closing ceremony[31]).

[28] The organizers of this *magal* claimed in 1984 that it was nine years old, having begun in 1976. There appears to have been an annual meeting since 1976, upgraded to pilgrimage status in 1980. See *Le Soleil*, 4 Sept. 1980. My source: interviews with Babacar Lo, a Saint-Louisian Mouride, London (May 1985).

[29] *Ndigel*, 13 (Mar.–Apr. 1984).

[30] Interview with Babacar Lo, London (May 1985).

[31] Quoted in *Le Soleil*, 1 Oct. 1984.

These young Mourides should be seen in part as the leaders of Senegal's Muslim fashion in the 1970s; Young Mouride associations were formed in the principal Senegalese towns, and were organized in a National Federation in 1978. A Mouride youth culture emerged, with a new style of dressing (the Ahad *boubou*) and of talking (some new words). The brotherhood's magazine, *Ndigel*, saw much more than this—a Mouride style in painting, in literature, and in architecture: 'Mouridism has achieved the dimensions of a veritable civilization. The Mouride population is young and highly intellectual.'[32] The cultural identity thus proclaimed is very much an anti-Western affirmation; as in the reported words of the President of the Federation of Mouride Youth: 'We must rescue our Muslim youth from the darkness of cultural alienation.'[33] Young children, specifically, should be sent to Muslim schools rather than to the *école maternelle*.

The intellectuals and the organized young, following their restatement of Mouride faith in Amadu Bamba, 'the Man of Refusal', have also asserted a certain independence from the brotherhood's traditional leadership. They begin with a critique of the brotherhood's country magicians, the *marabouts à cartes*, those who boast that whisky turns miraculously to milk in their stomachs. They continue by questioning the claim of Amadu Bamba's family to be the destined heirs to his spiritual gift. 'We are not praise-singers for the M'backé family,' they say, 'we are the spiritual sons of Amadu Bamba. You don't have to be an M'backé to be a spiritual guide.'[34] Such declarations are little short of revolutionary within the Mouride brotherhood. The title *shaikh* has in the past been the sure index of Mouride power, and it has gone with blood descent (from the founder or one of his associates). Modern Mouride intellectuals prefer the title '*guide spirituel*' to that of *shaikh*. The *shaikh* may be the object of prostrate veneration for the rural disciples, but to these urbanites he seems to be little more than a country bumpkin.

The task of the spiritual guide is of course firstly one of pious

[32] *Ndigel*, 13 (Mar.–Apr. 1984).

[33] Interview with Mustapha Diop, reporting his conversations with the leaders of the Mouride student association, Paris (Apr. 1985).

[34] Interview with Doudou Diop, President of the Mouvement Islamique des Mourides d'Europe, and with other members of the MIME committee, Paris (Apr. 1985).

instruction, to teach the disciples to be good Muslims as well as good Mourides. This is a task which follows directly from the work of the founder: 'Amadu Bamba's teaching is no more than an extension of the Prophet's teaching in the city of Medina.'[35] The text of the Qur'ān is supplemented by the founder's poetry (his *qaṣā'id*) and by a study of the prophetic *sunna*. This is, in short, a movement of Muslim reform in the Mouride style, away from the tradition of blind obedience to the *shaikh*, towards a more modern form of Islamic devotion which might have a greater claim to international Islamic respect. There may be a problem here in that the linguistic medium of reform is not Arabic but either Wolof or French; but this is not the only movement of Islamic reform in which a European language has an important function.

A Mouride reform movement also suggests a power struggle, of as yet uncertain outcome, within the Senegalese Islamic community. Senegal's Islamic reformers have since the 1950s been critical of certain malpractices associated with the Mouride brotherhood (such as popular saint-worship, neglect of Islamic education, and magical practices). Informed local observers can thus discern in Senegal's 'Muslim reform' a disguised attack from the local Tijāniyya on the Mouride brotherhood. A Mouride reform movement prepares the brotherhood's defence against attack from this quarter, while suggesting the counter-attiack. Amadu Bamba, the Man of Refusal, not only protects Islam from the assult of the West, he also refuses subordination to the Arab world—unlike the Senegalese branch-leaders of the Tijāniyya or the Qādiriyya, deferring as they do to Fez or to Baghdad.

Mouride defence of African cultural pride is a powerful attraction to the young. Here lie some intriguing possibilities for the brotherhood's further growth, as has recently been seen in the confrontation with the Tijāniyya in the Senegalese cities, with the frequent conversion of Tijāni youth (attracted by the racial and national appeal of Amadu Bamba, also by the Mouride reputation for superior organization and power—'if you want to get a state scholarship, you'd do better to be a Mouride'). The brotherhood's popularity among the young is

[35] Abdoulaye Dièye, quoted in Diop, 'Les Associations murid et la Da'wa en France', p. 6.

tellingly shown in the changing fashions of popular music, where reggae singers substitute 'Bamba' for the Rasta chorus 'Jah Jah'.[36] It was even reported that the reggae star Jimmy Cliff had become a Mouride, following a recent visit to Touba. And in Paris there is a West Indian ex-weightlifter (a former European champion) who has become a Mouride *guide spirituel* and who hopes to bring the Mouride brotherhood to the West Indies.[37] Amadu Bamba in such cases stands for black racial pride as well as being the bearer of an Islamic identity.

The point of convergence for all these various Mouride energies is of course the symbolic figure of Amadu Bamba, and then the real figure of the founder's successor, the Khalifa-General, Abdou Lahat M'backé. Although the Mouride Khalifa is more than adequately supplied in luxuries (a Cadillac, a Rolls-Royce, recently his own airstrip), his political position is not without its discomforts. First there are the brotherhood's traders, upon whom he has grown increasingly dependent, who have shown themselves willing to act in an independent manner. The Khalifa's instructions have not commanded an automatic compliance from these businessmen, either in the long-running scandal of the use of Touba as a headquarters for contraband, or, for example, in the relative defiance of his instruction to vote for Abdou Diouf in the 1983 general election.[38] The traders do in their way remain good Mourides, giving lavishly enough to the holy aristocrats of Touba, but they have a disturbing inclination to make up their own minds.

And then there are those tricky intellectuals with their endless talk of their devotion to Amadu Bamba (or, as Mourides call him, Serigne Touba). These intellectuals insist that their obedience goes exclusively to the Khalifa-General (he, after all,

[36] Interview on the Mouride musicians with Mar Fall of the Centre d'Étude d'Afrique Noire, Université de Bordeaux (Apr. 1985).

[37] Interview with Amadou Guy Pépin, Mouride spiritual guide and ex-weightlifter, Paris (Apr. 1985).

[38] The governing party of course won everywhere in the Senegalese parliamentary elections of February 1983, but, in so far as a Mouride share of the vote can be disentangled from the regional results, it seems that the governing party did less well in Mouride areas, where abstention was relatively high, as was the opposition party vote—this despite the Mouride Khalifa's instructions to vote for the government candidate. See D. B. Cruise O'Brien, 'Les Elections sénégalaises de février 27, 1983', *Politique africaine*, 11 (Sept. 1983).

represents Serigne Touba); and their international activity certainly helps to make Serigne Touba better known, and thus helps to increase the Khalifa's power. But can they be trusted? The Khalifa clearly had some reservations when he asked an intellectual delegation: 'Are your intentions pure?'[39] He appears to have answered this question almost ten years later in a communiqué on the subject of the Association Nationale des Jeunes Mourides du Sénégal (29 January 1986), where he warns the administration and the public against this association, which has 'moved away from the teaching of Serigne Touba' with its 'programme which would include the whole country, later perhaps all of Africa'. The association is here bluntly condemned for indulging in 'practices contrary to Mouride faith, to honesty, and to moral probity . . . As representative of Serigne Touba, I must condemn these things which damage our brotherhood and undermine its greatness.'[40]

These young Mouride intellectuals have scarcely troubled to hide their scorn for the unlearned traditional *shaikh*, with his crude magic and his country clientele. And the Khalifa-General, although by background a country *shaikh* himself (of limited education), has also shown his irritation with the lazy and luxury-loving aristocracy of Touba. But the devotion of the disciple to his *shaikh*, in this world and the next, is at the base of the edifice of Mouride power—unpaid labour, cash offerings, and political obedience all flow with that submission. The Khalifa-General gets a major share of these tributes; he could not welcome any attack on the *shaikh* from within the Mouride brotherhood. And it was the founder himself who warned that the disciple without a *shaikh* would find Satan for his guide.

[39] Interview with Saliou Mboup, Dakar (Jan. 1981).
[40] 'Communiqué du Khalife', in *Le Soleil*, 29 Jan. 1986.

# 8

# Charismatic Authority and the Creation of a New Order
## The Case of the Mahdiyyat Movement in South-Western Nigeria

PETER B. CLARKE

COMPARED with research on Islam in other parts of West Africa, scant attention has been given to the subject of charismatic authority and millennialism or Mahdism in south-western Nigierian Islam—this despite the fact that, on occasion at least, both have been major ingredients in Muslim belief and community formation in that region. The Mahdiyyat movement treated here is one of the best documented cases of Islamic millennialism in south-western Nigeria, and is of interest not only because it throws light on the scope and force of this belief among south-western Nigerian Muslims, but also because it is of that relatively rare type of Islamic millennialism, the pacifist variety.[1]

The millennialism propagated by the Mahdiyyat movement, in a society torn asunder by political strife and experiencing severe economic hardship, functioned largely as an ideology for change, by focusing attention on what were perceived to be unprecedented changes occurring in 'these last days' (the Second World War being an unmistakeable sign of the end of time) and by promising a profound moral transformation which would make all things new. An examination of the emergence of this movement provides an opportunity for an analysis of the character and style of charismatic authority and responsiveness found in south-western Nigerian Islam, which, it should be noted, has developed in close proximity to and has been influenced by African Independent Church Christianity.

[1] This account is based on field work in Nigeria, in 1975–8, 1981, and 1982.

## 1. THE PATH FROM REFORMER TO MESSIAH

The Mahdiyyat movement was formally launched in Ijebu-Ode, the capital of Ijebuland in south-western Nigeria, on 20 December 1941 by a prominent Ijebu Muslim, Muhammad Jumat Imam (1896–1960). On that day, in conformity with local tradition, which required anyone with a new religious idea or acrobatic or magical display to make this known first to the local ruler and his council, and in this way obtain a licence or permit, Muhammad Jumat Imam went to the palace of the Awujale (King) to announce to the ruler in council and to those who had assembled to hear him that he was a prophet of God. More precisely, supporting his claim with Qur'anic and Biblical quotations, he proclaimed himself to be the expected Mahdī (divinely guided one) of the Muslims and the promised Messiah of the Christians. The title 'Mahdī-Messiah' which he claimed was obviously borrowed from the Ahmadiyya movement, which was active in Ijebu at the time and which at one point, according to informants, Muhammad Jumat Imam had seriously considered joining.

If his claim was acknowledged and his message received by the ruler and his people then, so the Mahdī-Messiah informed the Awujale, God would make the Ijebus his chosen people, transfer his blessings from the Arabs and the Jews to them, make Ijebu-Ode his holy city, remove from it all bitterness, strife, and hardship, restore its freedom, and turn it into a centre of pilgrimage for people from all over the world. Rejection of his claim and message would meet with God's wrath and punishment and would see these divine blessings conferred on the town of Ile-Ife, the legendary centre and ancestral home of the Yoruba race, and his own mother's birthplace.

Muhammad Jumat Imam received his licence to preach (the Awujale's response will be discussed below), and on 2 January 1942, accompanied by his second wife, whom he presented as 'Shaikha Mahdī', and holding a Bible in one hand and a Qur'ān in the other, he proceeded to address a gathering outside the gates of the town. He informed those who had assembled to hear him that he had been 'inspired' to deliver his message there because the town had been deserted by the Spirit of God, on account of the evil practices of its inhabitants. Once again he proclaimed himself to be the Mahdī-Messiah, and asked his listeners to believe in God and in Jesus and Muhammad as his prophets. His mission was, he announced, 'to establish universal peace and to show Muslims, Christians, and pagans the only way to salvation'; through him God would bestow on Nigeria 'a glory and renown surpassing that of Mecca and Jerusalem'.[2]

There were various reactions to this declaration, as will be shown, but what is clear is that Muhammmad Jumat Imam was attempting to gain a following by transcending the numerous and serious divisions which lay within the Muslim and Christian communities, between Muslims and Christians, and between these two communities and the followers of the traditional religion; all of these divisions had been widened by the political turmoil that marked Ijebu society at the time.

This was not just the activity of an opportunist attempting to gain a following, for, as the account of his life given below indicates, he had been preoccupied with the issues of Muslim–Christian integration and Nigerian cultural and political autonomy for a long time before making this declaration. The division within the Muslim community was essentially over the question of the appointment of a Chief Imam of the central mosque, different factions advancing the claims of their own candidates. For some time Muhammad Jumat Imam appeared to stand aloof from the dispute, never interfering directly in support of one side or the other. However, he cleverly and indirectly made it known to his followers, of whom even before his proclamation there was a considerable number, that he believed that neither of the two main contenders was suitable, on the grounds that they were intellectually ill-equipped for the position. What counted in the Mahdī-Messiah's opinion was not only good character but also something that both candidates lacked—a knowledge of Arabic and English. That such a view met with the support of many is an indication of how both the expansion of Western education and the improvements in Islamic education (attributable to, among other organizations, the Ansar-ud-Deen society and the Islamic Reformation Society founded by the Mahdī in the 1930s) were altering the traditional criteria for leadership. Learning was becoming

[2] National Archives Ibadan, Nigeria (henceforth NAI), Ije Prof. 2 C 138.

almost as important as age, seniority, and good character, once the main and virtually the sole requirements for leadership. It is worth noting here that it is very often in situations such as this, where two value systems are in competition, that charismatic and innovative persons are successful.

The Mahdī-Messiah, widely considered to be a learned man, was not without ambition. The documentary evidence suggests, as do some of the oral data, that he was anxious to become Chief Imam himself, and that his failure to do so explains in large measure his decision to proclaim himself the Mahdī-Messiah. There is also evidence that he was prepared to join the Ahmadiyya movement on the condition that he was made the leader of his local branch. Many of his admirers, of course, could not see him in this light, for he was to them a prophet and therefore above the mundane pursuits and interests which preoccupied ordinary mortals. Furthermore, much of the evidence available to the historian, for example that contained in the Mahdī-Messiah's notebooks and diaries, where he reveals his ambitions, reactions, hopes, aspirations, and his sources of information and inspiration, was not known to his followers at the time. Consequently their appreciation of his actions and interpretation of his motives will, for this and other reasons, inevitably differ from that provided by the historian of today.

Understood retrospectively, in the light of the documentary and oral evidence available, as is the historian's way of understanding, there is an inevitability about Muhammad Jumat Imam's declaration. It is clear from his diaries, the records of his dreams and visions, and other sources that he was not only ambitious but also abnormally egocentric, constructing and arranging the world and interpreting signs and events so that all centred on himself. The day on which he was born, a Friday (the exact date is not known) in December 1896, the house in which he was born and its location, the year of his first pilgrimage to Mecca (1926), and numerous other happenings were all endowed by him with supernatural significance. For example, writing in 1941 of his 1926 pilgrimage to Mecca, he stated, 'I travelled around the world to London, France, and Italy, but I did not land in Italy, to Port Said, Alexandria, and Cairo, Jerusalem and Galilee. God showed me all those countries, but he forbade me to go to Germany or to land in

Italy. In his explanation this year 1941 He told me why he forbade me to go to Germany and Italy, there were enemies of his in each.'[3]

Moreover, although he was seen by many as genuinely charismatic, and was undoubtedly a man of presence with considerable personal charisma, there was more than just an element of the charlatan in Muhammad Jumat Imam, and there was a considerable amount in his charismatic performance that was pure fabrication. He would make clandestine visits to, among others, local Christian 'prophets' from the *aladura* (praying) churches, diviners, and numerologists, drawing on their advice and techniques and weaving this into his teaching and preaching as though it were part of a divine revelation of which he was the sole recipient. The principal divination technique used by him was that known in Yoruba as *yisabi* (in Arabic: *ḥisāb*), a system of astrological divination which determines signs of the zodiac by a mathematical process and not by the date of one's birth. But even when it was known or suspected that he practised divination it did little or nothing to weaken the faith that followers placed in him; for their perception of their leader was very much influenced by the ways in which he responded to their hopes and aspirations. What might well in Yoruba society have occasioned the withdrawal of trust were such things as exaggerated and unrestrained eccentricity, or especially madness, traditional symptoms of which were public displays of long hair, nakedness, and talking aloud to oneself. None of these was ever attributed to the Mahdī-Messiah even by his strongest critics.

Muhammed Jumat Imam was probably educated by his father Hassan Amokeoja, who had started a Qur'anic school in the Idepo quarter of Ijebu-Ode in 1904. According to a police report the father of the future Mahdī-Messiah was 'a native of Ijebu . . . a *mallam* [*teacher*] who had a thorough knowledge of the Qur'ān . . . [and] was said to be a troublesome man in the town to the extent that he was made Chief Imam of Ijebu-Ode but was later removed from office . . .'.[4] It may well have been his membership of the Tijāniyya, which was accused of

[3] University of Ibadan Library (henceforth UIL) MF 53 PK 12, Private Papers of Muhammad Jumat Imam, the Shaikh Mahdī of Ijebu-Ode.

[4] NAI, Ije Prof. 2 C 138.

spreading 'devil worship' and was fiercely opposed by the Muslim community in Ijebu-Ode, that cost him the imamate.

According to oral sources Hassan Amokeoja, who was the son of the Balogun (Minister of War) of Ijebuland, was one of the very first converts to Islam in Ijebu-Ode, and his children were among the first Ijebu Muslim children to be born of Muslim parents. His wife Zainafo Adefobe was, as we have noted, from Ile-Ife, and was said by Muhammad Jumat Imam to be of the royal family of that town. She worked as a petty trader while Hassan Amokeoja, in addition to his work as a Muslim teacher, farmed and bought and sold land.

Little is known about Muhammad Jumat Imam's childhood and youth. Although he apparently received no formal Western education he was fluent in both spoken and written English—something that was to amaze his followers, who would later attribute this ability, his numerological skills, and in general the 'wisdom', powers of understanding, and insight he displayed to the fact that he had received the 'gift of the Spirit of Truth'. In 1918 he went to Kano, settling in the ward of Kofar Mata, where he traded and worked as a *tafsīr mallam* to the Yoruba community until 1922, when he returned to Ijebu-Ode. It is very probable that while in Kano Muhammad Jumat Imam came into contact with Mahdist ideas there. Many Madhists from Zinder had migrated to Kano after the First World War, and a considerable number of those already there had joined the Tijāniyya, of which Muhammad Jumat Imam must by then have been aware, if only through his father's involvement in the movement.[5] He himself was later to join the local Ijebu branch of this Muslim brotherhood.

The next major influence on the Mahdī-Messiah appears, at least in retrospect, to have been his pilgrimage to Mecca in 1926. He was one of the first Ijebus to perform the *ḥājj*; this gave him considerable status in the local community and provided him with a platform in the shape of the Society of Alhajis which he founded in 1927. From this time until 1941, when he proclaimed himself to be the Mahdī-Messiah, Muhammad Jumat Imam was seen as an ardent reformer of Islam. He established the

[5] See J. N. Paden, *Religion and Political Culture in Kano* (Berkeley: University of California Press, 1973) 169–71 and *passim.*

Islamic Reformation Society for this purpose, and helped to set up and actually taught in an Islamic school with an integrated Islamic and Western curriculum. He also taught Arabic in the school founded in 1904 by his father Hassan Amokeoja.

The vision many people had of the Mahdī-Messiah was of a reformer and not of an ambitious, power-hungry, status-conscious individual. Moreover, there was little in his teaching and behaviour that could cause offence or alarm. The vast majority of those living in the Idepo quarter of Ijebu, where the Mahdī-Messiah preached and taught, were Muslims. A growing number of these, while anxious to provide their children with a Western education, feared that by sending them to schools run by Christians they would endanger their Muslim faith. They were also anxious to change the image of the Muslim as 'bush' and uncivilized—that was current among Christians in Ijebu at the time. Islam was widely regarded as little more than 'paganism' in a new guise—a view of local Islam shared in fact by Muhammad Jumat Imam, who insisted that the faith and practice of Islam was not incompatible with Western education, science, and technology, and that as long as Muslims implied that it was they would continue to be dismissed as backward and primitive. Muhammad Jumat Imam was also concerned with the high level of illiteracy in Arabic, maintaining that this was in large measure responsible for the 'idolatrous' behaviour and practice of many of those whom he labelled purely 'nominal' Muslims.

It was also his opinion that Muslims tended to 'deify' the Prophet Muhammad, who, he explained, was simply a prophet, indeed no more than one among many prophets chosen by God, all of whom were equal. Interestingly (in the light of his declarations of 1941–2), no mention was made of Muhammad as the last of the prophets. Furthermore, he told Muslims they were deceiving themselves by thinking, as many apparently did, that Muhammad could intercede for them and 'save' them, or that the pilgrimage to Mecca guaranteed entry to paradise, or that the Qur'ān contained a superior form of revelation to that contained in the Bible; both were God's revealed word and therefore of equal value.

His views on Muslim women and worship aroused considerable interest. He preached not only that he was acting in accordance with the Prophet Muhammad's practice in educating Muslim women, but also that they had every right to attend the mosque and worship alongside men, something that was disallowed in Ijebu mosques at the time. There was nothing in the Qur'ān, he maintained, to forbid this, nor was there anything in the Qur'ān to suggest that women should be in purdah. This brought him considerable support from Muslim women, as did the announcement that God had declared his wife to be Shaikha Mahdī, to act as his lieutenant and his eventual successor.

By the mid 1930s Muhammad Jumat Imam had become a leader of some stature in Ijebu-Ode, and a representative and defender of the interests of a growing number of Muslims. He was referred to by the Acting Resident of Ijebu Province, in a 1935 letter to the Secretary to the Western Provinces, as 'a leading member of the Muhammadan community'.[6] Muhammad Jumat Imam had on this occasion, as on a number of others, successfully petitioned the colonial authorities to give assistance to a member of an Ijebu Muslim family who was stranded in Jedda, on the return journey of the pilgrimage to Mecca. He took up many other 'causes' with the colonial authorities, relating to taxation, land reform, and religion, and came to be seen not only as a reformer and modernizer but also as a highly skilled and successful mediator.

He also developed close ties, as mentioned above, with a number of Christian 'prophets' of the *aladura* (praying) churches, which were beginning to have (and continue to have) a profound impact on the religious life of Ijebu society, and indeed on that of Nigeria and West Africa as a whole. In fact he lived no more than 300 metres from St Saviour's Anglican Church, where Pastor Shadare worshipped, who was one of the founders of the *aladura* movement and well known to the Mahdī-Messiah. He also went for advice and counsel to Prophet Olagbami, a member of the Christ Apostolic Church of Ijebu-Ode, who interpreted many of the Mahdī-Messiah's dreams.[7]

Although he had risen considerably in stature and impor-

[6] NAI, 4 J 803/1 (Pilgrimage to Mecca); Acting Resident, Ijebu-Ode, to the Secretary, Western Provinces, Enugu (6 Aug. 1935).

[7] Interview with Prophet Olagbami, Ibadan (28 Feb. 1978).

tance, and had acquired a reputation as a learned man among Muslims and in the wider society of Ijebu-Ode, Muhammad Jumat Imam did not allow his style or standard of living to alter unduly. He continued to live in the one-storey house in Idepo Street where he had been born—two storeys being a sign of prosperity and success. He wore traditional Yoruba dress—frequently abandoned by the successful for European clothing—and continued to teach, farm, and trade in books. Although a critic of much in his own society and a champion of reform, Muhammad Jumat Imam, unlike many of the Western-educated élite, but like some of the extremely influential prophets of the *aladura* churches, remained very much a part of Ijebu society; he kept abreast of new ideas in education, politics, and religion, and of new moods and trends, without allowing them completely to reshape his own way of life and behaviour.

However, it is clear from his own writingas that by 1941, the year of his first proclamation, Muhammad Jumat Imam had begun to attach far greater significance to his dreams and visions than he had before. For example, he recalled that in 1936 a voice from God had commanded him to write an epistle to all the Muslim Emirs in Nigeria, warning them to 'follow more faithfully in the path of Allah'.[8] The emirs of Ilorin and Zaria replied, and the latter allegedly sent a certain Shaikh Bab Ahmad to express his approval of Muhammad Jumat Imam's opinions. The twelfth of December 1938 was the occasion for another celestial visitor, this time a spirit under the orders of St Michael, who informed him that he had been chosen by God 'to give a message to all Muslims'. Henceforth, Muhammad Jumat Imam wrote, 'spirits used to come to me every night until the appearance of the Holy Spirit of Truth on August 29th, 1941.' In 1939 a senior, prominent, and respected Muslim in Ijebu-Ode, al-Hajj Abdulkadir Eleri, recounted to Muhammad Jumat Imam a dream of his own in which he saw the Holy Spirit descending on the latter and raising him up to preach to the whole world. Then in 1941 Muhammad Jumat Imam was told by two Christians from the Faith Tabernacle Church, situated near his home in Ijebu-Ode, that 'God was prepared to give him

[8] NAI, Ije Prof. 2 C 138, Government Correspondence concerning Shaikh Mahdī of Ijebu Ode.

a crown if he would devote himself entirely to his service'.[9] It is worth noting, in the light of his declaration in December 1941 and his announcement of the purpose of his mission (the unity of Christians and Muslims), that the 'messengers' from God came from both the Christian and the Muslim faiths.

But this was not the only evidence that Muhammad Jumat Imam was called to be a Messiah. Numerology was brought in to provide further proof that he had been divinely chosen for this role. That he had been born on a Friday and had entered Mecca on a Friday was, he explained, of great significance; for Friday corresponded with Jumat, his own name, and 'Jumat' and 'Messiah' in its Arabic form, '*masīḥ*', possessed the same numerical value of 118—clear evidence that he had been destined to be the Messiah.[10]

By 1940 he had already begun to make what were considered by many at the time to be 'prophetic' pronouncements—for example, that the Europeans would soon be 'driven out', allowing Nigerians to rule themselves, and that the price of *gari* would increase astronomically. There was in this, and in much of what he said and did, a strong element of cultural nationalism; indeed his insistence on using Yoruba alongside Arabic in the mosque was largely motivated by his nationalist feeling. To argue that Arabic alone was God's language was in his opinion a form of cultural and linguistic imperialism. Furthermore, his relegation of the role of the Prophet Muhammad, referred to above, and his decision to proclaim himself a prophet were to some extent prompted by this same consideration: to demonstrate that Africans were as worthy as Arabs or Jews or any other race to be prophets of God.

## 2. THE SOCIAL CONTEXT OF CHARISMATIC AUTHORITY

The foregoing provides the bare bones of Muhammad Jumat Imam's life up to December 1941, when he decided to proclaim

[9] UIL, Private Papers of Muhammad Jumat Iman.

[10] Ibid. (It may be objected that this numerological evidence provided by the Mahdī rests, according to Maghribi reckoning, upon a misspelling of *masīḥ* as *maṣīḥ*.)

himself the Mahdī-Messiah; while it sheds light on his character, intentions, and personality, it does not fully explain the relatively wide acceptance of his charismatic authority, or the appeal of his millenarian message. For such an explanation we have to look to the political and religious turmoil and economic hardship in Ijebuland at the time, and the very great fears aroused by the Second World War. The size of his following must not be exaggerated; it was never more than 30,000 at any one time, out of a total Ijebu population of around 80,000—though his influence was probably greater than these figures might suggest.

When he decided to advance his claim to be the Mahdī-Messiah Muhammad Jumat Imam stepped into a situation of virtual anomie. A society that less than fifty years previously had been one of the most xenophobic, closed, tightly-knit, and self-sufficient in southern Nigeria, with a strong sense of its own direction, had been prised open and carved up by numerous competing interest-groups.[11] When the British took control of Ijebuland in 1892 there was hardly a Christian or a Muslim to be found there, for successive Awujales had refused to allow missionaries of either faith to proselytize for fear that they would disturb the unity of their kingdom. But by the 1940s the vast majority of the inhabitants of Ijebu-Ode, Ijebu-Igbo, Shagamu, and the other major towns of Ijebuland had been converted to either Christianity or Islam.

By this time also the authority of the Awujale, the traditional ruler of Ijebuland, had lost all legitimacy; this was due ultimately to the introduction of the 'Sole Native Authority System' in 1917. The position of Awujale rested on the principle of divinely legitimated inheritance through three specific 'houses' or families. It was, according to a local historian, 'an object of reverence, the repository of the collective conscience and spiritual development of the Ijebu people'.[12] The authority of the Awujale was defined by law and custom, and his power was shared by various institutions, which prevented any individual or group from becoming all-powerful. Executive power rested in the hands of a body of titled officials known as

[11] See E. A. Ayandele, 'The Ideological Ferment in Ijebuland, 1892–1943', in id., *Nigerian Historical Studies* (London: Frank Cass, 1979), pp. 270 ff.

[12] Ibid., pp. 280–1.

the *Ilamuren*. An organization known as the *Regberegbe* (Age-Grades) was the guardian of the rights and powers of the people. According to the same historian it had the power to 'alter the decisions of the four societies, the *Ilamuren*, the Odi, the Osugbo and the Pampa' and was composed of 'all the "natives" of Ijebu'; 'the Awujale and chiefs derived their titles from it . . . its decisions were final.'[13]

This type of constitutional monarchy was swept away with the introduction of the 'Sole Native Authority System', in which the Awujale was appointed by the colonial government and made subordinate only to the Resident of Ijebu Province. However, until 1929 the British continued to choose the Awujale from one of the three 'houses' or families and thus respected the principle of legitimacy. Moreover, the Awujales continued to rule in the traditional way. But with the appointment in 1933 of a Western-educated Awujale, who was determined to make full use of the powers vested in him by the 'Sole Native Authority System', and who according to many was 'illegitimate', coming as he did from the once-defunct Gbelegbuwa 'house', the opposition began in earnest. It reached a climax with the 'Monster Petition' of protest signed by 2,635 taxpayers and presented to the colonial authorities in 1943.[14] The period from 1933 to 1949, when the 'Sole Native Authority System' was abandoned, saw assassination attempts on the Awujale, who sought permission to carry a revolver, and the polluting of the water supplies with juju. Riots, often led by women carrying the *akoko* leaf, a symbol of protest, and chanting 'the crown has fallen off your head', were common in the early 1940s. But even more subversive was the symbolic rejection of the Awujale's authority by Muslims who abandoned the traditional practice of calling as a community at the Awujale's palace on their return from the Id prayer-ground, after the Lesser Beiram festival, to pray for him, pay him their respects, and show him their loyalty. This decision by Muslims to break with custom in such a serious way was announced in the Central Mosque, and the few who disobeyed were banned from worshipping there. A similar practice by the followers of the traditional religion was also discontinued; the latter, moreover,

[13] Ibid., p. 280.
[14] NAI, Ije Prof. 7/4, The Ijebu Monster Petition.

carried on their sacrificial rites and other ceremonies on important occasions without the Awujale being present, thereby seriously going against tradition and even endangering the validity and efficacy of these rites.

Opposition to the Awujale and to the 'Sole Native Authority System' did not serve to unite either the Muslims as a community or Ijebus as a people; in fact they both continued to be leaderless to all intents and purposes. Meanwhile the colonial authority, with the rise of nationalist politics, was itself coming under increasing pressure and lacked any legitimacy. In fact during the Second World War colonial rule was often compared with Nazism on the grounds that it deprived people of their legitimate right to govern themselves.

As we have seen, there were deep divisions in the Muslim community—over the appointment of the Chief Imam of the central mosque, and between the orthodox Muslim community and the 'heretical' Ahmadiyya movement, which had arrived in Ijebu from Lagos in 1929, and also between certain groups of Muslims and the Tijāniyya, which was accused of promoting 'devil worship'. Education was one of the issues that divided Muslims and Christians; the former accused the latter of using the schools to convert Muslim children, while the Christians themselves had splintered into numerous different groups, some of which, for example the Watch Tower Movement, not only publicly condemned Muslims but also assigned other Christian denominations to hell. Moreover, among Christians there was as we have noted a tendency to despise Muslims as backward and illiterate.

Very little in society seemed to function well. Even Western education and medicine were found wanting. The former was charged on the one hand with creating a new generation of insubordinate and arrogant youth, and on the other with producing unemployment; while the hospitals were overcrowded, lacked equipment and staff, and seemed incapable of satisfying even the most basic of people's expectations in the sphere of health.

Independence politics had begun in an organized, formal way in Ijebu in the 1930s and were dominated by the Nigerian Youth Movement (NYM), founded in 1936. Here again there was division and confusion as young and old, Christian leaders and Muslim leaders, the Western-educated and the upholders of traditional rule competed for the right to represent the interests of the people. Moreover, there was considerable disagreement over how far the independence issue should be taken in the context of the Second World War. If the single-minded, dedicated pursuit of this goal were to undermine the war effort then, one group argued, the consequences could be disastrous, leading to Nazi domination of Nigeria—a fate far worse than British rule. It was for this reason that this group withheld support for the independence movement in India. Many people were aware of Hitler's contempt for the black race; anti-German propaganda abounded in Nigeria, and many believed that if Britain fell they would be subjected to his tyrannical rule. Others were for pressing ahead with all the means available to achieve the goal of independence.

The Second World War profoundly influenced the thinking and attitudes of the Ijebus and others in Nigeria, and added an extra urgency and significance to the millenarian theme which provided one of the few remaining untarnished ideologies of change. During the war years many Ijebus believed along with other Nigerians that they were living in a time of acute political, social, and moral crisis which heralded the Last Days. They had seen many of their institutions, political, religious, social, and economic, challenged and replaced by foreign ones brought in by colonialism. They had experienced the questioning and rejection of some of their long-standing assumptions about stability and normality, and witnessed the traditional sources of authority being undermined by representatives of those very same powers which were now tearing the world apart and threatening a universal catastrophe. The war indicated clearly to many that the 'civilized' nations which had taken control of their destiny, and that of Nigeria and Africa as a whole, were themselves in need of transformation, and that what they represented in terms of human and spiritual values was profoundly suspect.

People had been placed in a situation in which they could neither retreat to the traditional order nor have confidence in the present order. There was much bewilderment and confusion, as depicted in the words of one local prophet: 'On the Last Day the heavenly host will say, "Welcome, welcome from the peril of Vanity Fair indulged in by those who forget their Father's house" . . . the souls of those who safely went through the confusion of the world will dwell with the Lord Jesus.'[15] People searched both for an explanation of the war and for a new path forward. Many believed that 'sin' had caused the war and that only through a process of moral and spiritual regeneration could it be brought to an end, making way for the millennium, an age of peace and justice, which would precede the end of time.

Serious economic problems also coincided with the war years, leading to a steep increase in the price of basic foodstuffs such as *gari* at a time when people were being requested to make sacrifices and to contribute to the war effort. The war affected life at every level, disrupting families, providing women with an unprecedented (and to many heads of families unacceptable) degree of independence and freedom, and drawing attention to rights and privileges denied to Africans by Europeans in South Africa and elsewhere.

## 3. THE RESPONSE TO CHARISMATIC AUTHORITY

It was in this context of confusion and disunity that the Mahdī-Messiah launched his Mahdiyyat movement. Under the Madhī-Messiah's authority and that of his second wife, Shaikha Mahdī, were a number of *khalīfa*s who were made responsible for the movement in the various towns where it had been established, among them Ilorin, Ibadan, Abeokuta, Epe, and Ijebu-Igbo. Like Tijāniyya and other Muslim brotherhoods the Mahdiyyat movement had its own *wird* or special prayer which was recited at the beginning and end of every meeting and act of worship and many times during the course of the day. It was basically the Muslim confession of faith with the addition: 'and Muhammad Jumat Imam is the Mahdī-Messiah'.

In 1944 the Mahdī-Messiah opened the newly built Mahdiyyat mosque which, in line with his mission to unite Christians and Muslims, had been constructed in the form of a mosque and a church and was presented by him as 'the mosque which all

[15] NAI, Oyo Prof. 162, Commissioner of Police's copy of 'Awon Asotele'.

Muslims and the temple which all Christians have been expecting at the end of the world'.[16] This was to symbolize the New Jerusalem where Muslims and Christians would no longer quarrel, for there they would discover 'that both worshipped the same God in the way they chose'.[17] And to demonstrate even further his ecumenical spirit and reforming and modernizing zeal the Mahdī-Messiah employed the Christian rite and form of marriage and used blessed water and oil (an *aladura* rite) for the purpose of spiritual healing. Initiation into the Mahdiyyat movement itself followed the pattern of the Christian rite of confirmation. He insisted, moreover, that all his followers study the Bible and the Qur'ān since they were both God's word and therefore of equal value.

This 'New Jerusalem' which the Mahdī-Messiah was in the process of constructing was to be founded on tolerance, not only that tolerance which ought to exist between Muslims and Christians, but one that accepted that black people had the right to their own prophets and were worthy to be chosen by God for such a role. He was fully aware that for this level of tolerance to become the norm a moral revolution would be required, or, as he expressed it, 'a *jihād* (struggle) of the heart and mind'.[18] It was, however, something that could be realized in practice.

As to the social composition of the Mahdiyyat community, teachers and businessmen, journalists and nationalist politicians, and senior members of the Muslim community all featured in the movement. The majority of his followers, however, consisted of the semi-skilled and members of the poorer classes: tailors, bricklayers, washerwomen, fishermen, and carpenters. By 1959, the year of the Mahdī's death, the movement had over 20,000 members and had established centres in all the major cities in Ijebuland and beyond. But the death of the leader and the succession of his junior wife Shaikha Mahdi saw charisma turning graceless, for these events provided the occasion for a bitter dispute among members and the emergence of a breakaway group led by the senior son of his first wife and his supporters, who rejected the right of Shaikha

[16] NAI, Ije Prof. 2 C 138.
[17] Ibid.
[18] Ibid.

Mahdī and her sons to lead the movement. This dispute was eventually resolved on the death of Shaikha Mahdī, and today the Mahdī-Messiah's tomb in the Mahdiyyat central mosque is a place of pilgrimage for all Madhists.

As has been mentioned already, the response to the Mahdī-Messiah's declaration and mission varied enormously; we can now consider this very broadly, from the standpoint of the political and religious authorities on the one hand, and from that of the ruled or governed on the other.

### *The Response of the Colonial Administration*

The colonial administration's response to Muhammad Jumat Imam and his Mahdiyyat movement was in large measure shaped by a concern to safeguard the principle of religious liberty, but at the same time avert any possibility of a backlash from orthodox Muslims, locally or in northern Nigeria. The administration therefore moved quickly after the declarations of December 1941 and January 1942 to put the self-proclaimed Messiah under surveillance, directing the Commissioner of Police of the Western Provinces to open a file on him. In the biographical section of this report Muhammad Jumat Imam is described as 'an astronomer and trader who owns a shop dealing mostly in Arabic books'; the report continues: 'he had the order of the *khalifat* [*sic* for *khilāfa*] conferred upon him by Shehu Umaru of Hadejia in 1937, and in the same year was made the *muqaddam* of the Tijāniyya in Ijebu. The Chief Imām opposes his activities accusing him of attempting to create a division among Muslims because he was not chosen as Chief Imām.'[19] According to another report, from the Resident of Ijebu Province to the Secretary to the Western Provinces, the Mahdī-Messiah was 'popularly regarded as an eccentric and is treated as such by Muslims and others'.[20]

The administration took a rather more serious view of the situation when it discovered that the Mahdī-Messiah had decided to take his message to northern Nigeria; the Secretary to

[19] Ibid., Assistant Superintendent, Nigerian Police, to the Resident, Ijebu Province (10 Dec. 1945).

[20] Ibid., Acting Resident, Ijebu Province, to the Secretary, the Western Provinces (24 Dec. 1945).

the Western Provinces advised the Resident of Ijebu that 'The Mahdi and Promised Messiah ought to be persuaded to desist from prophesying and made to realize that a Muslim messiah who quotes the New Testament for his purposes is unlikely to be taken seriously by Muhammedans in the Northern Provinces.'[21] The Mahdī-Messiah's letter, or *wathīqat* as he called it, to the Emirs of the northern provinces has already been mentioned; even more disturbing for the administration was his arrival in Kano and later Maidugari *en route* to Mecca at the head of some five hundred followers, a group of whom were carrying the Black Box, which was claimed to be the Ark of the Covenant containing divinely revealed and sacred objects. Further uneasiness was felt when it became known that the Mahdī-Messiah had met with Sir Sayyid Abdel Rahman Pasha in Khartoum. This prompted the Secretary to the Northern Provinces to write to the Chief Secretary to the Government at Lagos, warning that 'Such fanatics as the self-proclaimed Mahdī-Messiah should not be allowed to travel where they gather a reputation which they may trade on when they return home.'[22]

After the Khartoum meeting the government files increasingly depicted Muhammad Jumat Imam as suffering from mental derangement, an opinion they alleged to be shared by the general public. The Secretary to the Western Provinces and the Resident of Ijebu Province both agreed in 1946 that, in the words of the latter, 'Though a fine figure of a man and of some personality he is not in his right mind.'[23] They were very much mistaken in thinking that most people agreed with their diagnosis of the Mahdī-Messiah's mental state.

### *The Response of the Traditional Authorities*

The Awujale and his chiefs, while opposed to the Mahdī-Messiah's activities, argued that as long as the principle of religious toleration was regarded by the colonial government as

[21] Ibid., Secretary, Western Provinces, to the Resident, Ijebu Province (5 Dec. 1945).

[22] Ibid., Secretary, Northern Provinces, Kaduna, to the Chief Secretary to the Government (23 Nov. 1945).

[23] Ibid., Resident, Ijebu Province, to the Secretary, the Western Provinces (10 Dec. 1945).

'an article of faith' there was no alternative but to allow Muhammad Jumat Imam to establish his movement, and to treat the whole affair as a matter for the local Muslim community.

The Mahdī-Messiah was both a threat and an embarrassment to the local ruler, and on several occasions attempts were made to prosecute him. The grounds were 'disrespect' and 'insulting behaviour' towards the Awujale—which consisted of, among other things, the Mahdī-Messiah's use of a large, wide, brightly coloured umbrella, a symbol of the Awujale's status and authority, and his refusal to recognize and show the customary respect to the Awujale by bowing low in his presence; this practice was condemned by the Mahdī-Messiah as non-Islamic. Indeed one of the more interesting aspects of Mahdism in south-western Nigeria is that it raised again in this Muslim outpost a number of contentious issues of a doctrinal and ritual kind that had once been a source of controversy, both among Muslims themselves and between Muslims and non-Muslims, elsewhere in West Africa. The question of bowing or prostrating before a ruler, for example, takes one back through the centuries to early Islam in the royal courts of the western Sudan, where people stood before the king 'like one performing a *rak'a*', a practice that our source, Ibn Battuta, ridiculed then and one that successive Muslim reformers were later to condemn.[24]

Reform was in the air among Muslims in western Nigeria in the 1930s and 1940s, and in the writings of the *'ulamā'* of the region one can discern the beginnings of an attempt to construct an Islam with its own distinctive and separate identity. To this end such questions were raised as that of whether Muslims should be buried at ground level, in the manner of the traditional religions, or below or above it. Muslims were warned against taking part in politics because they might have to compromise their beliefs by carrying out the directives of, and associating with, non-Muslim rulers. And polytheism was frequently attacked, a variety of arguments being advanced to prove the oneness and uniqueness of God.[25] However, it is worth

[24] Ibn Battuta, *Travels in Africa and Asia, 1325–54*, sel. and trans. by H. A. R. Gibb (London: Routledge and Kegan Paul, 1929), 327.

[25] See A. K. Bidmus, 'A Literary Appraisal of the Arabic Writings of the Yoruba *Ulama*', M.Phil. thesis (University of Ibadan, 1972).

noting that, while he presented himself as a Muslim reformer, the Mahdī-Messiah's attitude towards the Awujale was an example of the way in which the reform of Islam could be used for political ends—in this case to attack the 'Sole Native Authority System'.

Traditional priests also felt threatened by the Mahdī-Messiah, who proceeded to make a public display of burning the amulets and 'idols' of 'pagan' converts to his movement. On one such occasion in 1949 the traditional priests protested to the Awujale, accusing the Mahdī-Messiah of disdaining their gods, annoying the spirits of their ancestors, and profaning the land. The Mahdī-Messiah was tried, found guilty, and given the choice of one of three punishments: banishment, imprisonment, or a heavy fine. The Mahdī-Messiah sought to use the verdict as further proof of his claim to prophethood by informing the Awujale that he would accept any of the punishments in good faith, since all genuine prophets to date had been ill-treated and persecuted by their own people. Anxious at all costs to avoid turning the Mahdī-Messiah into a martyr, the Awujale shifted responsibility to the local courts; but here again, and for similar reasons, the judges were reluctant to impose a sentence, and the defendant was subsequently acquitted and discharged, providing his followers with further proof that their leader was no ordinary mortal.

### *The Response of Christian and Muslim Leaders*

A certain degree of hostility to the Mahdī-Messiah from Christian leaders was inevitable, for he had presented himself as their Messiah and claimed to have a fuller knowledge and understanding of the Old and New Testaments than they did. The local Anglican bishop dismissed the Mahdī-Messiah as a false prophet. On the other hand 'prophets' of the *aladura* churches tended to be more sympathetic and positive. Many of them had been voicing similar criticisms of colonialism, and predicting a similar fate for it. Prophecy 149 on the list of prophecies confiscated by the colonial government predicted: 'Something is coming down into the black peoples' land to the extent that all white people who live in it will perish.'[26] And the

[26] NAI Oyo Prof. 162, 'Awon Asotele . . .'.

millennarian theme was a persistent one: 'The Lord Jesus is coming soon', prophesied Prophet Abiodun Babatunde in 1930, and Pastor Odubanjo preached: 'We are at the end of the end . . . the present age will soon give way to the next, the Golden Age of the Millennium.'[27] The Watch Tower Movement's preaching in Ijebu-Ode was equally millennarian creating an atmosphere of both gloom and expectancy.

There was, therefore, much in common between the Mahdiyyat movement and a number of Christian movements, which made for a certain tolerance and mutual support. This was not so with the Muslim leadership, the strongest opposition coming from the two rivals for the position of Chief Imam, whom the Mahdī-Messiah regarded as unfit for leadership on the grounds that their lack of knowledge of Arabic and their generally low level of education disqualified them from holding this office. He stated that 'Before a man can become Chief Imam he must be a good Muslim, free from all diseases, well-educated in Arabic, know more than others, know how to translate the Qur'ān for the people . . . and because of this knowledge he will be able to advise and govern, and even command the Holy Spirit.'[28] This emphasis on the knowledge that comes from book-learning, making it, rather than the wisdom that comes with age, the most important prerequisite for leadership, gave rise to considerable tension between the Mahdī-Messiah and the Muslim leadership in general, which was largely illiterate in Arabic, Yoruba, and English.

The Ahmadis deeply resented the fact that both the title and the role of their founder, Ghulam Ahmad, had been usurped by Muhammad Jumat Imam, and challenged the latter to prove that he was in fact the Mahdī-Messiah by 'producing fire from his belly in the midst of an assembled crowd'. Witnesses apparently counted to ten but no fire appeared, so the Ahmadis called Muhammad Jumat Imam 'a liar, to his face' and went on to ask him: 'How can there be two Mahdīs, yourself and Ghulam Ahmad, and how can the Mahdī come from Ijebu-Ode?'[29]

[27] Pastor Odubanjo, *The End of the Age . . . Christ is coming soon* (Lagos, 1947), 21; Prophet Abiodun Babatunde's words are from NAI, Yoruba News, 16–23 Sept. 1930.

[28] NAI Ije Prof. 2 C 138.

[29] Interview with Mr Kuku, who founded the Ahmadiyya in Ijebu-Ode in 1929 (22 Oct. 1977).

While a majority of the Tijāniyya refused to follow Muhammad Jumat Imam after and because he had declared himself the Mahdī-Messiah, they felt little hostility towards him, continuing to regard him as 'a good leader who knew both Arabic and religion well and who made many good reforms but who unfortunately allowed his frustration over not being made Chief Imam to cloud his judgement.' In fact the Tijānis had testimony from Sharif Ahmad, who had given Muhammad Jumat Imam his *silsila*, that his former pupil could not possibly be the Mahdī.[30]

The colonial administration apart, there was never any suggestion from other leaders that Muhammad Jumat Imam was mentally disturbed, and, while most of them explained his behaviour in terms of the frustration caused by his failure to become Chief Imam, few, if any, denied that he was a man of considerable ability and charisma.

### *The Popular Response*

We can now turn from the 'official' response to a consideration of the ways in which ordinary people, and in particular Muslims, reacted to the claims made by Muhammad Jumat Imam. This has already been touched upon; once again we have a variation in response, ranging from complete acceptance to total rejection, with qualified acceptance and indifference characterizing the response of quite a number.

In characterizing a response as one of complete acceptance our intention is not to suggest that it was automatic and unquestioning, but rather to indicate that after being convinced by rational and moral argument, and by 'supernatural' signs and 'miracles'—for example many believed the Mahdī-Messiah's claim that he could communicate directly with the dead, and saw this as a 'supernatural' power—people decided to commit themselves to his guidance. Many followers were also impressed by the 'scriptural' evidence he produced in support of his claims, recounting how when they approached the Mahdī-Messiah and asked him to verify these claims he did so by reference to the Bible and the Qur'ān. However, it was perhaps

[30] Interview with al-Hajj Kassim of the Tijāniyya, Ijebu-Ode (28 Oct. 1977).

the Mahdī-Messiah's numerological skills as much as anything else which convinced enquirers, who explained: 'It was his power to group words and to deduce from them such things as the name of God, of the prophet Muhammad, and his own name, Jumat, which he wrote down and put together to mean Messiah.'[31]

Others, not knowing his background in any great detail, were persuaded that since he had received no formal education he must have been the recipient of a 'divine gift of knowledge' which enabled him not only to 'group numbers' but also to predict the future. Being so learned without a formal education put him, it was believed, on a par with the Prophet Muhammad. In the words of one informant, 'The Mahdī-Messiah never went to school yet he forecast many things. How could he know and understand so many things except for the fact that he was chosen by God and possessed the Spirit of Truth'. Those who accompanied him to Mecca in 1945 were impressed by his ability to answer the questions put to him by the *'ulamā'* of Kano and Borno. 'All their questions he answered,' one informant explained, 'yet he does not go to school to learn to read. That is what convinced us. From where, we ask ourselves, does he get his knowledge?' Strange as it may seem, the idea of the self-educated man does not appear to have occurred to them.

The Mahdī-Messiah's gift of prophecy was, as we have seen, acknowledged as a divine gift even by devout Christians, including Mr Otufale, a pillar of the Christ Apostolic Church of Ijebu-Ode, who stated: 'The Mahdī told me my future in 1936 and it has come true. I do not believe in prophecy or fortune-telling so much, but an individual can know that this Mahdī was rightly guided by God.' Local market women, among others, recalled the Mahdī-Messiah's forecast that there would be a sharp increase in the prices of *gari* and rice, and spoke of it as a 'marvellous prediction'.

But it was not only the 'miraculous' and 'supernatural' side that attracted people; 'good character' and 'true religion' were also an integral part of the Mahdī-Messiah's appeal. Women in

[31] This section on the popular response is based on interviews with both members and others who refused to join the movement, conducted during the periods mentioned in n. 1.

particular laid much emphasis on these elements and found in him 'a champion of our rights in religious matters'. They welcomed the 'innovations' which allowed them to worship in mosques alongside men and to leave their homes and go about their business during the Oro festival (a traditional festival from which they were excluded by custom and during which they were confined to their compounds). The majority of women who joined the Mahdiyyat movement were illiterate and engaged for the most part in petty trading. They refused to abandon Muhammad Jumat Imam when he proclaimed himself to be the Mahdī-Messiah because, as one explained, 'He was a very religious man and of good character, a balanced man who directed people in the right way', and this was proof enough for them that he was 'anointed by God and a prophet of Allah.'

What was 'balanced' about the Mahdī-Messiah, in the opinion of these women, was the way in which he concerned himself with their position in society and the tolerance he showed to others, especially Christians. In the words of one woman follower, 'He preached the one way to Allah and belief in all holy books, that not to believe in the Bible as well as the Qur'ān is not to be a Muslim, and he gave women their equal position with men in society.' For others, he removed fear—fear of spirits, gods, the evil eye, dreams, and various other phenomena that troubled their lives.

As for the Mahdī-Messiah's 'gift' of counselling, this consisted above all else in his interpretative powers, especially where dreams and visions were concerned. 'If you went to him and were in trouble,' one informant recalled, 'you would have a dream in which you sinned, and you would go back to him to tell him, and he would then tell you what to do, and by the power of God you would overcome your trouble.'

The Mahdī-Messiah also attracted support as the 'black prophet' among a people made increasingly aware of the movement for political independence that was gathering momentum in southern Nigeria. Many gave positive support to the idea advanced by Muhammad Jumat Imam that since 'The Whites and the Reds [*Arabs*] had their own prophet, the Blacks must have their own too.[32] They also endorsed the view put

[32] UIL, Private Papers of Muhammad Jumat Imam.

forward by the Mahdī-Messiah that Islam was as much a Nigerian religion as an Arab one; this largely explains why he and his supporters disputed the finality of the prophethood of the Prophet Muhammad. To many of his followers this idea was part of a process of political and cultural domination, and they wanted with their leader to know why, 'if God has given Saudi Arabia its prophet and Palestine its prophet and other peoples their prophets, He cannot also give black people their own prophet; for are we not also made by God?'[33]

That the Mahdī-Messiah went unrecognized by many did not constitute a reason for disbelieving his claims, for it had been the lot of all genuine prophets to be despised, ignored, and rejected; in support of this view disciples cited the cases of Jesus and Muhammad. The fact that he was not widely known was also attributed to the absence of a written and widely spoken Nigerian language, and to the fact that the Mahdī-Messiah was black. Those knowledgeable in the Qur'ān would find evidence therein that Muhammad Jumat Imam was a prophet; for example, Qur'ān s. 9 v. 89 was interpreted by recourse to numerology: the passage was said to foretell that the Messiah of the Last Days would emerge in a place called Kada, which name was said to have the same numerical value as 'Ijebu-Ode'.

## 4. CONCLUSIONS

To those Ijebus, then, who followed Muhammad Jumat Imam there was ample evidence to confirm their faith and trust in his teaching and mission. He was the prophet of the Last Age called to unite Christians and Muslims, to slay the Antichrist, Hitler, and to restore to the Ijebus and all black people their liberty and independence; the result would be a New Jerusalem of peace and harmony. The expected millennium would be of this world and would come about through the 'cleansing and purification' of the present order, achieved through the correct understanding and practice of Christianity and Islam as taught by the Mahdī. Its advent was imminent, for the Mahdī-Messiah would

[33] The quotation is from an interview with a Nigerian soldier from Ijebu who heard about the Mahdī while on war service in Sierra Leone; it echoes the sentiments of many other followers of the Mahdī interviewed by the author.

very shortly slay the Antichrist and put an end to 'Satan's seed'. There would be no further strife in Ijebuland or Nigeria.

Here was an ideology of change for a society reacting to the turmoil generated by rapid developments which were undermining the cultural, social, moral, and political fabric of the old order. Some welcomed the changes that were taking place and went out to meet them; others opposed them, turning back to the past; while others tried to live with one foot in the past and one foot in the present. But even those who welcomed change, no less than those who responded indifferently or negatively to it, were often unable to face it in the raw, naked, form in which they encountered it.

What the Mahdī-Messiah's millenarian ideas and his movement appear to have provided was an ideology and structure which enabled many, in particular those who were aware of the need for change but who were also strongly attached to their own culture, to rework what they considered to be valuable traditions and customs and integrate these with the desirable aspects of the emerging new order. In this way the Mahdiyyat movement not only held out the promise of stability and order in a new society but also provided a means of digesting, emotionally, intellectually, and spiritually, the changes necessary to bring this about.

# 9

# Charisma and Medicine in Northern Nigeria

MURRAY LAST

## INTRODUCTION

OF the just and learned Ibrahim Dabo, forty-fifth ruler of Kano (1819–46), it is written: 'The prodigies Dabo performed and his miracles were many; I mention only a few for fear of exaggeration.' The miracles mentioned were all conquests of towns—and (a matter of apparent significance for the nineteenth-century chronicler to record) it was only after he had conquered a town that Dabo permitted himself to have his head shaved once more. The source of his success? At the start of his reign, he had said (as the chronicler narrates it), '"I will conquer [the rebels], if Allah wills". He entered his house and remained there forty days praying to Allah for victory. Allah heard his prayers. . . . he made a camp on Dalla hill . . . he spent many days on Dalla—perhaps forty, I am not sure—and then returned home'.[1]

The image the Kano Chronicler offers here is clear. Charismatic power, of the sort that works wonders and captures towns, is obtained through prayer—in this instance, prayer at the two poles of symbolic power: at the ancient palace of the great reforming king Rumfa (1463–99) and, over in the opposite half of the city, at a war camp on the hilltop once sacred to the city's oldest pagan cult. The important mosques that lay between these two poles—mosques whose importance had been made plain earlier in the Chronicle—are not mentioned as sites

[1] Kano Chronicle, translated by H. R. Palmer and published in volume III of his *Sudanese Memoirs* (Lagos: Government Printer, 1928), 128. For this chapter I have used various copies of the Arabic text (sometimes referred to as *Ta'rīkh arbāb . . . Kanō*) to offer a more accurate rendering of some of the quotations I give here. This study is part of a larger undertaking to produce an edition and new annotated translation of the Kano Chronicle. For an analysis of the Chronicle's composition, see my 'Historical Metaphors in the Kano Chronicle', *History in Africa*, 7 (1980), 161–78. Professional warriors (*jarumai*) always wore their hair long, whereas scholars had shaven heads; long hair was considered to look fierce (and wild?), as it was in Europe until very recently; cf. Richard Holmes, *Firing Line* (Harmondsworth: Penguin, 1987), 34.

for prayer in this context; they evidently symbolized another kind of power.[2]

But in fact the historical narrative is more ambiguous about the sources of power. The previous ruler, Sulayman (1806–19), also a scholar and the first of the new *jihād* Emirs, faced from the start the insubordination of his colleagues, and sent to Shaikh ʿUthman b. Fudi, the leader of the *jihād* and its most charismatic scholar, and asked his advice. The answer, so the account goes, came simply in the form of a sword and a knife; the chronicler adds a message, as if the meaning might not be clear—'to kill all who opposed him'. So prayer in this instance was considered less effective than simple violence.[3]

Finally, the Chronicle shows how to lose power. The ruler immediately before this, Muhammad al-Walī (1786–1806), had been driven out of office by the *jihād* forces and killed—an ironic fate for an Emir known as 'the Saint' (*al-Walī*), an Emir moreover who, we are specifically told, successfully saved his people from more mundane disasters like famine. His fault, so the chronicler suggests, was that in his attempt to purify the local practice of Islam he had caused the State's 200-year-old talismanic Qur'ān (wrapped in goats' skins and then cattle hides) to be opened up. For people had told the *walī*: ' "you will be removed from this city as surely as you remove that Qur'ān from its covering." And so it came to pass.' Indeed, power slipped away from him with the same sudden finality with which the white egrets in Katsina had burst out of their sacred, leather-covered clay building once it too had been opened up by a similarly doomed reforming ruler; then, 'the egrets filled the

[2] Kano Chronicle, p. 128. On Kano's clan mosques see M. S. Zahradeen, 'The Place of Mosques in the History of Kano', in B. M. Barkindo (ed.), *Studies in the History of Kano* (Ibadan, Heinemann, 1983), 57–66. The Chronicler is apparently portraying Dabo as seeking royal power (*sarauta*, associated with warriors and spirits) rather than religious power (*baraka*, associated with scholars and mosques), yet using the Muslim techniques of prayer to Allah and, perhaps, seclusion to do so, thus appropriating the royal sites for Islam.

[3] Kano Chronicle, p. 127. The added message is given in a marginal note to the Arabic MS used by Palmer. In the context it is significant to note that Shaikh ʿUthman never himself joined in the *jihād* fighting. Traditionally scholars did not carry arms. It is possible, as Dr Humphrey Fisher has pointed out, that the knife signified a razor for shaving new converts to Islam—though not, apparently, to the author of the marginal note.

whole town'. Thus it was in Kano: as secret knowledge became public, so power was lost.[4]

For nineteenth-century historians looking back to the momentous revolution that had occurred in their childhood, such transfers of power raised crucial questions about the *sources* of power. These sources, past and present, are the subject of this chapter. For charisma—outstanding power—is deemed to be not just earned (as are all gifts) but also manufactured and serviced. A long life, like a long reign, requires the right foods, the necessary tonics, to sustain it at full strength whatever the vagaries of the day. The recipe for power is seen to centre on a correct mix of raw force and that subtle ingredient which we variously label as confidence, faith, skill, magnetism, or awesomeness; the same ingredient which is now in Britain colloquially called 'bottle', and which makes a football star (as the fans say) 'walk on water'.

The science—or alchemy—of fusing the ingredients of power was, and still is, the speciality of Muslim scholars; and it is to their role in creating confidence and fortifying the charisma of leaders through the use of *ṭibb* (as that science is often called, translated usually as 'medicine') that I wish to draw attention here. At the same time, I wish to emphasize the essential ambiguity in people's attitudes towards this (artificial?) manufacture of charisma. Indeed tales of efforts, and final failure, to achieve either long life or a long reign are commonplace—and seem crucial to keeping charismatic pretensions to scale.

A central problem in societies where no group comes near to achieving a monopoly of violence is the relationship between physical violence and other forms of coercion. Charisma, in this context, is one means of compelling obedience without the use of physical violence; through it people are mobilized, apparently of their own free will. It is this ability to deceive people that is the peculiar property of power; just as force, through fear, can turn the living from persons into things, so power makes people's minds and wills mutate. The two means of control over others interact: whereas separately both force and charisma have their

[4] Kano Chronicle, p. 82 (Katsina); cf. pp. 116, 127 (Kano). Egrets, being all white and the habitual associates of cattle, are presumably here thought to symbolize Muslim Fulani scholars and pastoralists. Traditional cattle, however, were black, and their hides were used in non-Muslim ceremonials of kingship.

limits, together each potentiates the other, as in the synergism of drugs—force needs the confidence of charisma, charisma needs the ultimate sanction of force. The history of force (at least in its military manifestations) is tolerably well known, but a history of the special knowledges that underwrite local systems of charisma is more elusive.

I am using charisma here to denote force of personality in contrast to force of arms; together, as power, they ensure success. Medicine strengthens both; for, in the idiom of anatomy, the mediciner reinforces the heart and head, the eyes, ears, and mouth, the belly within or the skin without. These metaphors reflect the experience of us all in times of great stress when our psychology and physiology so obviously interact.

In this chapter I will take a single text, the Kano Chronicle, to illuminate from within a culture of power and the scepticism that characterizes it. The Chronicler and the subsequent contributors to the text were themselves scholars, writing in Arabic, but the events they recorded were mainly the actions of kings. They did not exclude successive kings' failures to ensure for Kano a military might commensurate with its wealth and status—for commercially Kano was considered in the sixteenth century to be on a par with Cairo and Fez.[5] More importantly, the authors of the Chronicle were apparently interested in the technological developments in this culture of power: the development of royal symbols, of military equipment, and of religious methods and skills. There was a market for these improvements and innovations, not least when they concerned the ways in which prayers might be produced. Competition, whether in prayer or other religious ritual, is not restricted to proponents of different faiths; almost as intense is the rivalry between sects (and individuals) within a single faith. Charisma,

[5] See D. Lange and S. Berthoud, 'L'Intérieur de l'Afrique occidentale d'après Giovanni Lorenzo Anania (xvi$^{e}$ siècle)', *Cahiers d'histoire mondiale*, 14/2 (1972), 338–9. During this period Kano and Gao were rivals in size; Gao was enumerated and found to have 7,626 houses, excluding stalk-shelters on the edge of town; see O. Houdas (ed. and trans.), *Ta'rīkh al-Fattash* (Paris: A-Maisonneuve, 1913), 145–6. From such a figure one might guess a population of some 75,000, of whom perhaps 10,000 would have been adult males. The population would fluctuate seasonally, with up to 50% being of slave status. An army which consisted only of free men aged 20–40 might only muster, say, 3000 men at most. In practice, campaigns might involve less than a hundred 'knights'—on one notable occasion, as few as 24; see O. Houdas (ed. and trans.), *Ta'rīkh es-Soudan* (Paris: A-Maisonneuve, 1900), 168–9.

then, is seen not as a single, universal attribute, but as a product always being improved to meet local requirements.

## 1. CHARISMATIC POWER AND MUSLIM SCHOLARS

Power, say scholars in Kano as elsewhere, is primarily from Allah. To be a friend (*walī*) or servant (*ʿabd*) of Allah is to have potential access to that power. Scholars, accepted into Allah's service after years of apprenticeship, have access not only to that power but also to the knowledge of how to use the powers that Allah has locked into his creations (plants, minerals, and animals; men, spirits, or simply words). Scholarship, then, whether expressed in textual learning or the regular recitation of prayer, is about both service and power. In theory, scholars should be invincible. Indeed they *are*—only Allah sometimes removes them from the scene of 'battle' (in their purity they have become so dear to him), thus giving us the illusion of their defeat and consequent death.[6] Allah, in short, is not as easily bent to men's will and their pursuit of power as people might suppose. For this reason other sources of power have to be explored, sources apparently under men's more immediate control. For Allah, or rather his servants, do not have a monopoly of power, at least in the short term, in this world.

That there are other sources of power—most notably the recently dead (ancestors), but also spirits of various kinds—was not disputed; nor that access to powers of this sort is sometimes inherited (as witchcraft), sometimes acquired (as sorcery). The serious dispute which scholars had with sceptics and non-Muslims centred on what might be the long-term consequences of someone making use of such powers. For scholars had to convince people of the reality of *bodily* resurrection, the day of judgement, and *eternal* life, none of which are self-evident even when the reality of ancestors and spirits is taken for granted; for Islamic eschatology is radically different from traditional understanding of what happens after death. In no way does it

[6] The model for Allah's intervention on behalf of his saints is of course Jesus, whom the Qur'ān (4: 157) says was not in fact crucified; Allah replaced him with a simulacrum. A similar substitution is popularly reported as having occurred at the time of the Sardauna's assassination (15 Jan. 1966).

conform to local common sense;[7] hence the central place that eschatology has in scholars' public teachings, with all the powers of poetry and eloquence called upon to quicken the imagination. Without Muslims taking seriously the belief in the day of judgement and the eternal, physical torments that might be meted out, the theological imperative to avoid resorting to the use of other powers than Allah's (especially his power as mediated by his servants the scholars) was scarcely compelling. Until people's scepticism on these issues was overcome Allah's servant was taken to be but one purveyor of power among many, and his claim to effective power (like any other such claim) had to be proven, here in daily life.

Muslim scholars asserted their difference from others in a number of ways, both political and symbolic. They wore, for example, cotton clothing in place of the more traditional leather; they might wear turbans, and keep their heads shaven. More importantly, as servants of Allah, scholars in West Africa traditionally did not carry arms or fight as soldiers for an earthly ruler. Nor did they pay taxes to such a ruler; to do so would imply serving someone other than Allah. Instead, gifts to Allah could be channelled through his servants—gifts of food, a bride, or lodgings. Juxtaposed to scholars within the 'culture of power' were the King's servants, who partook of the King's power, receiving free lodgings and a share of his taxes. In their service, of materially furthering the King's interests, they kept their hair long, were helmeted or hatless, and wore clothing and decoration derived from animals. Among the King's servants, however, the eunuchs were a special category: they alone had no other kin but the King who owned them, and they stood out, at least until reforms of the nineteenth-century *jihād*, as almost a secular equivalent of the scholar. Indeed, in the Kano Chronicler's mind eunuch and scholar come together (literally): 'in his [*Abu Bakr's*] time, there were many eunuchs in this land, and in his time too there were many scholars in the land; and the

[7] The debate, between 'pagans' on one hand and Muslims and Christians on the other, about the reality of bodily resurrection is very ancient. I have heard it in Nigeria, but it is reported from ancient Rome. If there is no day of judgement nor any threat of physical torment, what is there to stop one using magic? Eschatology is thus crucial to establishing practical if long-term, sanctions against such practices, essential as they are to traditional concepts of power.

land became too confined for the people.'[8] Both eunuch and scholar had surrendered their own physical power (the one as a father, the other as a soldier) in order to obtain another kind of power. Eunuchs were the King's *alter ego*: so identified were they with their King that they could impersonate him on the battlefield or in the mosque, and suffer martyrdom in foiling would-be assassins. Only scholars (and traditional priests) were expected to take service to such lengths, setting up their mystical power in counterpoint to raw force.

A proverb still current today in Kano remarks: 'the servant of the King—the remedy against Allah's servant' (*bawan sarki, maganin bawan Allah*). It is not just that 'Church and State' have conflicting interests; brute force has an immediacy that scholars cannot match. Yet the scholar can help make that force invulnerable, invincible: to stop himself being deposed a second time, King Kokona had his *imām* prepare medicine; set in the palace and in the houses of his two chief eunuchs, with fires constantly maintained over them, the shrines formed a triangle of power.[9] Similarly, military men mobilize all the sources of strength available, including that of the scholars. Sometimes the cost is very high, and not just the extra gold and silver that the *imām* demanded for his services to King Kokona. After the failure of local Borno scholars, a Moroccan *shaikh* was called upon to end a string of defeats, but (as ever with expatriate aid) the cost was extreme: the victorious king would have to die as part of the price.[10] Indeed, the Chronicler consistently notes the correlation between militarism and large-scale alms, for example: 'one could not find a king who gave as much wealth to *'ulamā'* as Kabe did (for paradise's sake and as alms), nor find one who slaughtered as many people.'[11] The fees which Allah's servants charged Muslim kings were a necessary expense in the maintenance of power. For without adequate medicine defeats could occur unexpectedly: 'in a twinkling of an eye Kano's

[8] Kano Chronicle, p. 114. Although a eunuch is first mentioned in the historical record some four centuries earlier (at Gao), eunuchs became a fashionable luxury only in 15th-c. Kano, with a special market established soon after for their sale. Castration is not an operation Muslims are permitted to perform; that, and the high mortality-rate of the operation (unless performed young), made eunuchs costly, and virtually a prerogative of royalty.

[9] Kano Chronicle, p. 121.

[10] Kano Chronicle, p. 116.

[11] Kano Chronicle, p. 125.

forces scattered because the [*enemy*] king had his poison with him.'[12]

Yet force required more than mere medicine; it required numbers, too—followers and allies. The Kano Chronicler readily attributed defeat to Kano's being outnumbered, and success on other occasions to the size of their host on horseback (as evidenced by the dry-season haze its dust created).[13] But numbers are not enough; steadfastness in battle, loyalty to the overall commander, unity on campaign—these are the ingredients of success, and they too are susceptible to medicine. The Prophet is quoted as saying 'war is trickery', and Ibn Khaldun asserted that 'victory in war as a rule is the result of imaginary psychological factors' and that numbers, weapons, and proper tactics, though important, are less effective in securing victory.[14]

A text like the Kano Chronicle is more than a listing of events, a source of disjointed data to be mined; it can be read as an extended commentary on sources of success and their efficacy (or failure). King after king is shown as seeking, often desperately, to stave off military defeat, to retain or extend political power; he and his advisers are shown as trying all the methods they know. Their search stimulated much innovation and experimentation, most of which has undoubtedly dropped from the record. What remains clear, however, is how fluid and contingent success, and the power to achieve it, were seen to be—and indeed were. For power in Kano never had in reality the structural stability which the ancient city walls and the great palace buildings sought to symbolize, and which was apparently attained by the Saifawa Mais ruling the neighbouring state of Borno.

## 2. THE CHANGING TECHNOLOGY OF CHARISMA

Developments in the market for power can be only broadly delineated. That Kano was constantly in need of ways to strengthen its resources of power is clear. As primarily a

[12] Kano Chronicle, p. 125.
[13] Kano Chronicle, pp. 115, 129.
[14] Ibn Khaldun, *The Muqaddimah*, trans. by F. Rosenthal, abridged edn. by N. J. Dawood (London: Routledge and Kegan Paul, 1967), pp. 229, 253.

merchants' city and a Muslim city, its government lacked the links to the countryside that would have given it access to reserves of manpower and skilled military leadership. Instead it relied on hiring soldiers, making alliances with marginal local groups, and on its wealth and contacts abroad to import military technology that would give its men an advantage over local adversaries.

Similarly, Kano's rulers had no intrinsic legitimacy of their own to focus the loyalty of the townspeople. The Kano Chronicle gives a continuity to the dynasty that other histories do not subscribe to.[15] Rulers were deposed and policies switched in a way that suggests factional politics dominated the office of kingship. The relative weakness of the ruler perhaps served the interests of the merchant community; they were ready to migrate to another city should taxation become in their view excessive.[16] In these circumstances, the rulers' search for power had to be acceptable to the various factions within the city, but what was acceptable was apparently also determined by whether it brought success.

As the Kano Chronicle portrays it, five phases can be discerned in the search for power:[17]

### *Phase 1* (c. *1350–1450*)*: Muslim Prayers and Presence*

Initially Muslim scholars (or merely educated Muslims) could publicly use prayer and parade themselves as an adjunct to the military tactics of their allies. The ritual was to walk in procession at night around the enemy's citadel and be present at the battle but leave the actual fighting to others. The only skill required was to *recite* by heart the Arabic prayers, and so be labelled as Muslim. On long-distance campaigns to difficult areas, however, it was not easy to persuade Muslims (whether merchants or scholars) to accompany the army; indeed, pious Muslim rulers became associated with periods of peace. As a

[15] For a more detailed attempt to reconstruct this period of Kano history see Murray Last, 'From Sultanate to Caliphate: Kano 1450–1800' in B. M. Barkindo (ed.) *Studies in the History of Kano* (Ibadan: Heinemann, 1983), 67–91.

[16] As happened under Kumbari (1731–43); see Kano Chronicle, p. 124.

[17] The Kano Chronicle's coverage of these phases is as follows: phase 1, pp. 104–10; phase 2, pp. 111–16; phase 3, pp. 117–27; phase 4, pp. 128–32. The analysis of phase 5 is not based on the Chronicle.

consequence, militant kings resorted once more to traditional prayers and rituals with considerable success—but not enough to prevent Kano becoming subordinate to the Muslim state of Borno and obliged to pay tribute. As the leading Islamic state in the region, Borno had real authority; its ruling dynasty claimed both Yemeni origin and descent from the *Quraish* of the Prophet. Kano's growing wealth, however, succeeded in attracting increasing numbers of Muslim visitors, and success for the city now apparently lay in fostering Islamic sources of power at the expense of traditional techniques.

But traditional techniques were not considered to be enough in themselves: parallel developments occurred in military technology. Armour for men and horses had to be introduced against the very effective poisoned arrows of southern armies; light cavalry were no longer adequate.

### *Phase 2 (*c. *1450–1600): Scholarship and the Sacred Book*

In this period knowledge of appropriate prayers, and the learning and libraries that validated them, increased greatly, as did the numbers of scholars. As a consequence there was competition between scholars for the top religious posts in the state—competition that was not always easy to resolve.[18] Technical superiority could be demonstrated in an academic 'contest by disputation', with quotation and counter-quotation. To be a visiting scholar, from Mali or Borno, was apparently an advantage, as was a North African education. By the mid-sixteenth century rulers of states could perform their own prayers, and did not depend solely on professionals; princes, too, were taught, and recited in public, the Holy Qur'ān. Earlier rulers had also emphasized their separate status by multiplying symbols of royal authority, acquiring enlarged walls and a palace, a huge harem and eunuchs, special trumpets and footwear. Militarily too there was innovation: spare horses for the king were now taken into battle, suggesting that the earlier safeguards had proved inadequate.

But once again all this proved ineffective against both internal coups and devastating external invasions; and once

[18] See *Aṣl al-Wangariyyin*, trans. in M. A. al-Hajj, 'A Seventeenth-Century Chronicle', *Kano Studies*, 1/4 (1968), 11, 13–14,

again resort was had to less orthodox defences as well. A talismanic Qur'ān and ark were installed, of a type used in Borno (where this had seemed to work well).

### *Phase 3 (*c. *1600–1800): The Age of Saints*

By the seventeenth century foreign scholars as a political force were *passé*. Local men had developed a subtler expertise, building permanent reputations as teachers, poets, and above all as saints. This first 'age of saints' may have been brief (its height was *c.* 1620–1720), but it left a lasting memorial. Whereas the earlier visiting scholars apparently seldom died in Hausaland and so left no focal point of sanctity, the saints' tombs have remained as places of pilgrimage to this day. Their graves also became focuses for other burials, as their blessing was sought in Barzakh; thus developed the practice of cemeteries in a culture where house-burial was the norm. Burial in the backyard of a private house effectively prevents a local pilgrimage (*ziyāra*) developing; with the creation of cemeteries the practice of prayer at holy gravesides, particularly by women seeking help in their private lives, was made easier. That this was a relatively late development, linked perhaps to the widespread adoption of Islam by women, is shown by the fact that though the earliest cemetery in Kano dates back to the fourteenth century, it has been built upon, and apparently was never associated with saints in the way of the other main cemeteries (known as *Waliyai*, 'Saints').

Sainthood was not confined to scholars. It was attributed to (or claimed by) certain rulers, though Kano sources do not agree on all the candidates. Outside Kano the most notable ruler was perhaps Mai ʿAli of Borno, known as 'the Cat' (a creature particularly beloved of the Prophet), who had powers unknown to more ordinary scholars: he could, for example, borrow books instantly from the shelves of al-Azhar without leaving Borno.[19] More importantly, he alone of his royal contemporaries successfully stalled the 'pagan' invasion of his state.

It seems that this age of saints is to be associated with the

[19] See J. E. Lavers, 'Diversions on a Journey', in Y. F. Hasan and P. Doornbos (eds.), *The Central Bilad al-Sudan* (Khartoum: El-Tammadon Press, 1977), 221.

development of Sufism in West Africa, and of brotherhood organization. Once again, as Sufi practice became commonplace, rulers had to become expert Sufis too. Thus kings performed the forty-day retreat (*khalwa*), spent in seclusion and prayer. The theory of power became ideologically more Islamic, and the kings in consequence reformist. But none of this saved the government from either military defeats or revolution. One famous saint met his death at the hands of invaders; another, protesting, disappeared, presumed to have been murdered at royal command.[20] One ruler drew up a final desperate battle-line, with ninety-nine scholars in front of him; few survived. Even the oldest traditional shrine, resurrected in desperation, was lost. Subsequently changes in military equipment were introduced: smaller cavalry shields, of the type used by southern cavalry, a bodyguard of gunmen (who had proved useful in Borno nearly two centuries earlier), and better fortifications for outlying towns.

### *Phase 4 (c. 1800–1900): The Young Sufi-Scholar-Soldier*

In both the sixteenth-century period of royal scholarship and the eighteenth-century period of royal Sufism, the rulers tended to be elderly. Wisdom, and the power it confers, were associated with age. The nineteenth-century *jihād* saw this association reversed: 'real' power, which combined both mystical and physical force, was now to be found only in the *young*. It was this new blend that gave such *jihād* leaders as Muhammad Bello in Sokoto or Ibrahim Dabo in Kano their unparalleled charisma.[21]

No one doubted the saintliness of the older leaders; indeed pieces of the hair of Shaikh ʿUthman were in great demand.[22] But the *jihād* saw the triumph of raw force as the instrument of mystical power. Death as a martyr in battle—an option not

[20] Not all the data here is from the Kano Chronicle. The death of one saint (Wali Mai Kalgo) is described in Abu Bakr Dokaji, *Kano Ta Dabo Cigari* (2nd edn.; Zaria: Gaskiya, 1978), 26; the murder of Abdullahi Suka is controversial (ibid., p. 28, but the mode of Suka's death is not mentioned). It was Bawa Jan Gworzo, Sarkin Gobir, who reputedly underwent *khalwa*.

[21] Muhammad Bello was 22 at the start of the *jihād* campaigns in 1804; Ibrahim Dabo was 28 when he became Emir in 1819, but he had previously led campaigns (for example, in Zaria) as Galadima.

[22] That is, in the Shaikh's lifetime; see H. Clapperton, *Journal of a Second Expedition* (London: J. Murray, 1829), 206. Subsequently his tomb has long been a place of pilgrimage. Bello's tomb has never been such a holy place, nor, of course, has Dabo's.

open to the earlier, unarmed scholars—reversed the taboo of sacred invulnerability which the Muslim scholar once sought. Prayer and militarism replaced the magic and mystery of the venerable saint or traditional medicine-man (*boka*). The focus on prayer and the separate but complementary emphasis on physical medicine for physical ills were part, it seems, of the programme of reforms for which the *jihād* was fought. It was, of course, made possible by the very youth of the *jihād* leadership, for whom the stress of battle and long campaigns was tolerable. But their charisma depended not just on youthful success, but also on their reputation for knowing the secrets of mystical power, secrets that sometimes not even their saintly fathers had known.[23] In short, mystical power could now be not only taught to the young, but also taken to levels previously unattainable. This was made possible by the fact that although great learning was an advantage it was no longer an essential qualification for these new levels of mystical power: whereas learning took a lifetime to acquire, mystical powers were attainable quite quickly given the right teaching and teachers. The Sufi brotherhoods were the instruments for establishing the new order of instruction, and they revolutionized the role of youth in West African Islam. Islam became a young man's faith, with the young apparently converting or reforming their fathers.

Against the armies of the colonial powers, manned as they were by West Africans recruited and trained by Christians (*Naṣārā*), the Muslim forces proved ineffective. Prayers were widely ordered to be said against the invaders, who were explicitly compared to Hulagu's Tartar hordes that overwhelmed the last Caliph in Baghdad.[24] Preparations were also made to defend Kano militarily. Stocks of gunpowder were held; the walls and gates, in good repair, were manned. Defeat, however, was evidently Allah's will.

### *Phase 5 (1900 to the present day): Scholars and Politicians*

During the first fifty years of the twentieth century, marked by a colonial monopoly of force and new economic opportunities for

[23] See Gidado b. Laima, *al-Kashf wa'l-Bayān* (MS 179, Ibadan), pp. 10–11.

[24] See Muhammad Bukhari b. Ahmad, *Risālat al-wazīr*, trans. in R. A. Adeleye, 'The Dilemma of the Wazir', *Journal of the Historical Society of Nigeria*, 4/2 (1968), 299–311. On Kano see A. Fika, *The Kano Civil War and British Over-Rule* (Ibadan: Oxford University Press, 1978).

the young, the power in Muslim circles swung back in favour of the elderly, in whom the custody of Islamic culture as a bastion against colonialism was vested. Despite the clear failure of both prayer and force to prevent foreign occupation, Islam's position as the community's ideology was, if anything, strengthened. But, since *c.* 1950, with the coming of the new politicians at the head of the campaign for independence, young men once again assumed political leadership. Simultaneously there arose a growing interest among the young in attending the sessions of prayer of the Sufi brotherhoods, as well as more formal legal and Qur'anic schooling.[25]

Lacking the established authority of high traditional office, many ambitious young men sought the support of Muslim scholars and their prayers, made the pilgrimage to Mecca several times, gave out alms, and financed the rebuilding of mosques and tombs. In the 25 years since independence, with a political career becoming increasingly hazardous, there seems to have been an increase in the money spent on prayers and similar products. The possibility of coups and similar political reversals of fortune, far from calling into question the efficacy of the techniques offered by scholars, has accentuated their importance, even among some of the military. This is nothing unusual, given, for example, Cabral's experience in Guinea-Bissao's wars of liberation, or the dependence of Marxist guerrillas in Zimbabwe on traditional healers.[26]

Following the oil boom of the 1970s, the funding available for religious activities grew considerably. Not all the funding of course has been local. Notable political figures of other African and Middle Eastern states, while mobilizing their own local sources of power, have reportedly also engaged certain Kano *shaikh*s to organize prayers on their behalf. The 'prayer economy' has, I suggest, come to make a significant contribution to Kano's prosperity. Indeed the apparent increase in the numbers of traditional Qur'anic students (*almajirai* or *gardawa*) in and around contemporary Kano is financed in part by such

[25] See J. N. Paden, *Religion and Political Culture in Kano* (Berkeley: University of California Press, 1973).

[26] See A. Cabral, *Unity and Struggle* (London: Heinemann, 1980), 58–60 (I am grateful to Sebastian Amanor for this reference); D. Lan, *Guns and Rain* (London: James Currey, 1985).

students finding employment as sayers of prayers, sub-contractors of the leading *shaikhs* who have been commissioned by politicians and merchants for so many thousand prayers. Such students can therefore also provide part-time unskilled labour at lower rates of pay than non-students; in consequence, as Paul Lubeck has shown, they have brought about changes in Kano's industrial economy.[27]

With prayer seen as an adjunct to success, not only can Muslim scholars readily find attachments to the households of the politically or commercially successful, but also notable independent *shaikhs* are once again courted for their blessings, through large-scale giving of alms. Such alms support the *shaikh*'s students and other followers, who can constitute a formidable bloc of support on the street. Not all have to be part of such established brotherhood networks as the Tijāniyya and Qādiriyya. Mai Tatsine, for example, was able to weld together a group of at least five thousand and take over parts of the city, before they were almost all killed in a series of bloody clashes in December 1980. Similarly *shaikhs* who are noted more for their powers of healing than for their teaching are keenly canvassed for their support; it is their medicine as much as their influence over the vote of their fellow constituents that is being sought.

Were the Kano Chronicler, then, to update his text today, he would find the city's streets full of stories about the mystical but politically practical powers of recently important figures like the Sardauna or the ex-Emir Sanussi. He would read in the newspapers reports of trials in which a governor explained how he had legitimately spent large sums from his security vote on buying prayers, or about a press conference in which a former Head of State suggested the use of juju against South Africa. He would hear the rumour that another Head of State, within days of taking office, sent up large sums of money for prayers to be said on his behalf by Kano's élite *shaikhs*. No doubt the Chronicler would dismiss as mere gossip the anecdotes about an impressive governor's waistcoat with its 'buttons of babies' eyes'—and how, when a courtesan stole it for a day or two, his power appeared to crack. He would hear how the preacher

[27] See P. Lubeck, *Islam and Urban Labor in Northern Nigeria* (Cambridge: Cambridge University Press, 1987); cf. I. A. Abba, 'Bara by Some Almajirai in Kano City in the 20th century', in Barkindo (ed.), *Studies in the History of Kano*, pp. 193–206.

mentioned above, Mai Tatsine, had been the *shaikh* behind an earlier, particularly robust and successful governor whose origins otherwise suggested that normally he could never wield such power. At the other end of the social scale, the Kano Chronicler might note people's interest (following an execution, televised live, of two armed-robbers) in what exactly was the medicine the robbers had used that had made it necessary for the firing-squad to shoot sixty bullets before the two were finally declared dead. Conventionally such gossip is ignored, but, as I have argued before, stories of this kind have real consequences; whatever their truth—and some undoubtedly are true—they are treated as true. They thus contribute to an individual's charisma, and create the reputations, generate the fears, and instil the confidence on the basis of which people make decisions and act.[28]

Over the centuries, then, the tools and techniques of Islamic power, administered through the agency of scholars, have been added to in the face of both failure and competition. At first prayers and the presence of Muslims were enough; then greater book-learning was required, and the eventual enshrining of a copy of the Qur'ān. After a series of disasters sanctity was seen to be embodied in the person of a saint, and subsequently in his tomb. The new Sufi experience and discipline made it possible for the young to embody sanctity; given a readiness to take up arms, the Muslim scholar was transformed into a ruler, militarily and politically active. In the twentieth century, with the monopoly of legitimate force out of their hands, politicians depended on popularity with a mass following. Once again, the Islamic idioms of sanctity were an essential element in the sense of power that popularity requires; but so too was medicine, particularly for those less able to attract a mass following or to protect themselves from rivals. For their problems were simply everyone's problems writ large; their solution lay in popular culture and the search for success.

28 Twenty years ago, in the heady days of Nigeria's first republic, such anecdotes were similarly rife. For a Sokoto perspective see Murray Last, 'A Note on Attitudes to the Supernatural in the Sokoto *jihād*', *Journal of the Historical Society of Nigeria*, 4/1 (Dec. 1967), 3–13.

## 3. ISLAMIC MEDICINE APPLIED TO POLITICS: THE CONTEXT TODAY

If, as I have suggested, medicine has long been an essential ingredient in creating charisma, and if, furthermore, charisma is essential for sustained power, then the problem remains of how it is that such cultural chemistry *still* proves so efficacious in popular politics. Mere continuity over the centuries ('tradition') is not enough. As part of the answer, I will outline some of the ways Islamic medicine is currently practised today—the argument being that normal, everyday practice of Islamic therapeutics lends crucial credibility to medicine used for political ends. In short, political medicine is not esoteric, but rather a logical, on a minor scale even commonplace, extension of medical practice, available to all who wish to afford it.[29]

The particular role of the Muslim scholar and the Islamic medicine (or *ṭibb*) discussed so far cannot therefore be properly understood outside the broader context of his daily practice. For a scholar's reputation is grounded in the respect which his neighbours have developed for him over the years; within his own community the efficacy and the limitations of a scholar's practice are transparent to all. Similarly, the success of a scholar in the wider political arena, and the patronage it may bring, feed back onto local demand.

I wish, then, to survey here Islamic medicine only in its strict sense: that is, therapies introduced by and identified with Islam, such as the use of Qur'anic texts and specifically Muslim forms of prayer. The range of treatments used by Muslims, whether in the Middle East, the Maghreb, or West Africa, is of course much wider than this, but much of the technology of Muslim medicine has been so long absorbed into local practice that it is no longer identified as 'Muslim'. By focusing instead on strictly Islamic medicine, one can identify the areas where Islamic power (*baraka*) is still considered peculiarly efficacious, despite the range of other therapies available in twentieth-century Nigeria.

[29] Much of the data in this section is based on fieldwork (1969–72) which focused specifically on medical practice in Hausaland. The more recent political data was collected when I was employed by Bayero University, Kano (1978–80), and on brief annual visits there (1984–6). I am indebted to very many colleagues and friends for their hospitality, patience, and teaching.

For Islamic medicine now has an important but specialized role within the wider Nigerian medical culture.

Islamic medicine is reserved for resolving primarily *social* problems. For success on a particular day, whether in trade, in politics, or in romance, the medicine of choice is a text from the Qur'ān. Popularity and the avoidance of enmities and of bad luck, these are available from scholars (or *mallams*), whose success-rate otherwise, with physical illness, is not so highly regarded. Qur'anic texts are either drunk (as *rubutu*) or worn as amulets (*laya*), but not all *mallams* supply such medicines for the market. Clients have their own favourite *mallams* whose *baraka* they believe in (or whose prices are right). Sometimes the only 'library' such a *mallam* has is scraps of paper on which appropriate formulae are written; *mallamin kundi* ('scrap scholar'?) is a term of contempt. Yet the same texts written by the very learned are much sought after. Conversely, my house has been used on occasions by students to mass-produce and bottle *rubutu*; and even I have been asked for, and have supplied, such medicines.

Just as *fiqh* orders the relations between individuals within the community, so Qur'anic medicine is the essential lubricant for day-to-day functioning. Like clothing or perfume, medicine transforms the raw social exterior of the individual into something acceptable. Food and drink sustain life; medicine sustains a good social life. And social life is defined as being centred on the towns and markets where Muslim rules apply.

It would be wrong then to interpret social relations in a Muslim environment as being exceptionally stressful, so that people need extra medicine merely to 'cope'. Its use reflects rather an awareness that the essential ingredient for success in a multi-cultural society is confidence; without it, success comes much harder in an alien environment where social finesse is at a premium. The ability to blend in with others is essential: the elaborate emphasis on humour and etiquette, the rules for negotiating, and the necessity for 'hosts', brokers, friends, or partners are all evidence of this. Seen in this light, Qur'anic medicine is truly a placebo; and the *mallam*'s *baraka* acts as the extra symbolism necessary to make it more than averagely successful.

This role for Islamic medicine is part of a wider underlying medical theory in northern Nigeria. According to this theory each group, each ecological or cultural zone, has its own diseases and its own cures. So the street-wise take the medicines of the block where they find themselves; Muslims in the deep countryside are advised to use local medicines even if pagan, while the countryman coming into town, or into a market run by Muslims, takes Muslim medicine. The Hausa farmer passing through bush needs Fulbe medicine; the pastoralist using farmed land for grazing will need farmers' medicine. Similarly, the 'modern' man requires modern medicines to do well in modern life; and Europeans have been synonymous with modernity.

Medicine, then, is primarily ecological, where ecology is the culture. In this context Islamic medicine is also seen as effective for Muslims' own special diseases, such as those of the eye, or piles, or even sexual disorders; Muslims are apparently considered especially prey to such problems—Fulbe notoriously so, but also *mallam*s generally (in contrast to the notably satisfactory 'strength' of the 'pagan' farmer). About a third of all entries in al-Suyuti's pharmacopoeia refer to semen and related subjects; the topic is far less symbolically or morally significant for Muslims than for Christians (masturbation, for example, is therapeutic; birth control, standard practice).[30] In the Nigerian context, then, Islamic medicine has a fund of 'recipes' not available elsewhere.

This special expertise may explain why occasionally some woman will announce that she has had a dream, and start healing a specific disorder with (Muslim) prayers. One such, an elderly woman named Amina Panteka, flourished for a few months before the local authorities closed her practice down; initially claiming to cure a non-specific women's ailment (*gishiri*), she expanded her practice to treat both men and women for a range of intractable complaints, for a fee (subsequently her practice flourished again, despite the initial ban). More often, women use saints' tombs as sites for pilgrimage and prayer, just as they use other naturally sacred

[30] *Ṭibb al-Nabī*, attributed to al-Suyuti, is translated in O. Elgood, '*Tibb-ul-Nabbi* or Medicine of the Prophet', *Osiris*, 14 (1962), 33–192. On Islamic attitudes to birth control before the 19th c. see B. S. Musallam, *Sex and Society in Islam* (Cambridge: Cambridge University Press, 1983).

places, in their search for cures for childlessness or loss of their children.

Islamic medicine is also preferred as protection against misfortune generally. Without necessarily implying that Allah is seen as the ultimate cause or as a 'High God', it is nonetheless considered appropriate to invoke his help, without excluding the use of other medicines for coping with more proximate causes. Thus rooms, houses, and cities require Muslim medicines for their protection, as do all people at risk (such as children), but the rituals or materials involved are often not strictly Islamic.

The communal use of Islamic medicine takes the form of ritual. When drought or an epidemic disrupts the community, the Emir leads a procession outside the town, with prayers followed by a sacrifice, the meat being distributed as alms. Usually, some days earlier, an attempt is made to 'clean up' the town by forcing the unmarried to marry or leave town, thus making the community in a sense ritually pure before the prayers are said.

Awareness that social corruption may result in natural disasters for the community is apparently common, but rarely acted upon. Some in Nigeria today go so far as to argue that the swing to fundamentalism should be understood as a conscious response to the austerity and economic failure which Allah has imposed upon the nation for its indiscipline. But the extent to which this self-consciousness about the moral health of the community as a whole could be the basis for a viable, long-term political programme is not so clear. Medicine is seen as intrinsically a short-term solution to an immediate problem; Qur'anic medicine preserves, it does not change a life-style, individual or communal. In this it is like ritual purity—a highly important state that permits one to be usually impure, without penalty. The *baraka* of both the medicine and the *mallam* is one mechanism that makes impurity permissible.

The *mallams*' major source of *baraka* or power is Sufism, both in its training and in the knowledge that it imparts. As specialists in that power, they can offer their services to politicians in their quest for power and to officials seeking to retain power. But Sufism is only one of a set of competing sources; the Arabic texts of recipes for political power verge in Kano on the merely magical, whilst protective charmed clothing, which derives its force from traditional ideas is on offer to politicians, industrialists, and thieves alike. The army of course has always been a major consumer of protective medicine of various kinds. Many of Shaikh ʿUthman's and Bello's miracles (*karāmāt*) related to battle, as did their opponents', or more recently Mai Tatsine's; for Sufism has highly practical uses, and need not only serve moral ends. Thus the suspicion among the legally-minded *fuqahā'* is that Sufis are merely seeking power, for themselves or for their patrons, over the community of Muslims by means external to that community; but if the community is truly governed by the *sharīʿa*, say the *fuqahā'*, such a power is unnecessary.

As the new Muslim near-equivalent of *mayyu* (sorcerers), such Sufi *mallams* can be the focus of all the odium reserved in the past for witchcraft—more so, one suspects, than those scholars who were involved in numerology or astrology—thus calling into question the very legitimacy of Islamic medicine as used to-day. So fundamentalist Muslims seek secular medicine, to undermine not just the Christian link to modern medicine but also the hold of Islamic medicine on the popular marketing of 'success' potions.[31]

For today's fundamentalists one of the major problems lies in the epidemiology of sickness. The majority of serious, potentially fatal illness is experienced by children and their mothers. By contrast, other adults have developed a relative immunity; and grown men, less exposed to the range of children's infections and better inured to the diseases of the wider world in which they operate, suffer least of all. Traditional medicine, whether Hausa or Islamic, has little success in curing children; and a medical system that regards illness as normal, and therefore merely counsels acceptance without trying to 'explain' the causes of illness further, is inadequate for a mother—however much it may satisfy men. Inevitably Muslim women have opened up space for their own therapeutic or divinatory system quite independent of Islamic medicine of any kind.

Given its rituals and scope for constant elaboration, the spirit-possession cult known as *bori* is more efficacious as a therapy than any Islamic treatment can be in relieving the problems and

[31] See, for example, Shaikh Abu Bakr Gumi, *Tarjamar Ma'anonin Alkur'ani Mai Girma* (Beirut: Dar al-Arabia, 1982), 715, note to sura 22, verse 17), where saint-venerating Muslims are equated with polytheists.

neuroses of women in purdah. For similar reasons *bori* is the treatment of choice in psychiatric cases—male or female—and even for epilepsy. Thus the officially sanctioned revival in Nigeria of interest, at both government and university level, in traditional therapies has threatened to legitimize spirit-possession cults on the grounds that they work, just as the television stations, in their search for local music, have popularized the cults' songs and their performers. As a consequence, the main debate in Muslim circles is now phrased in terms of 'culture' versus 'religion'. To this debate medicine, in one guise or another, is central; and on its outcome depend attitudes to *baraka*—or charisma—in modern Muslim culture.

## 4. CONCLUSION

I have suggested in this chapter that charisma, along with force, is seen as an essential component of success, and that writers like the Kano Chronicler are intrigued by the ambiguities that underlie extraordinary personal power, and by the way success eludes people. Given the role of scholars as counsellors, success was of course for them a matter of central concern, first in their competition with traditional, non-Muslim counsellors, and secondly in their relations with the military. I have also argued that the Kano Chronicler presents us with a history of how the scholars' inventory of techniques expanded as conditions required, how they not so much replaced old methods as supplemented and elaborated them. Historically only an outline of these developments can be reconstructed. The reader might be tempted to ignore such data as trivia from the past—justifiably so, were it not for the apparently expanding industry of prayer and Islamic medicine evident today. This expansion is but part of a wider search for success, linked not so much to austerity as to opportunity, a product of boom, not recession. Success in these conditions is not 'luck'; nor is charisma a chance 'gift', but an essential ingredient in an individual's ability to exercise power. To those short on charisma scholars offer a remedy; with it, people go about their affairs with that much greater confidence. Charisma, of a sort, is available to all.

# Bibliography

This bibliography has been compiled to reflect the fact that the volume is a Franco-British undertaking, and it also seeks to emphasize material on areas that have received relatively little attention to date. More detailed references are given in footnotes to chapters above, and in the bibliographical guides of P. E. Ofori (1977) and S. M. Zoghby (1978).


ABBA, I. A., 'Bara by Some Almajirai in Kano City in the 20th Century', in B. M. Barkindo (ed.), *Studies in the History of Kano* (Ibadan: Heinemann, 1983).

'ABD ALLAH b. MUHAMMAD FUDI, *Tazyīn al-Waraqāt*, ed. and trans. from the Arabic by M. Hiskett (Ibadan: Ibadan University Press, 1963).

ABUN NASR, J., *The Tijaniyya: A Sufi Order in the Modern World* (London: Oxford University Press, 1965).

ADELEYE, R. A., 'The Dilemma of the Wazir', *Journal of the Historical Society of Nigeria*, 4/2 (1968), 285–311.

AJAYI, J. F. and CROWDER, M. (eds.) *History of West Africa*, 2 vols. (3rd rev. edn.; London: Longman, 1985).

AL-HAJJ, M. A. 'A Seventeenth-Century Chronicle', *Kano Studies*, 1/4 (1968), 7–42.

ANDEZIAN, S., 'Pratiques féminines de l'Islam en France', *Archives de sciences sociales des religions*, 55/1 (1983), 53–66.

AUDOUIN, J. AND DANIEL, R., *L'Islam en Haute-Volta à l'époque précoloniale* (Abidjan: INADES, 1975).

AYANDELE, E. A., 'The Ideological Ferment in Ijebuland, 1892–1943', in E. A. Ayandele, *Nigerian Historical Studies* (London: Frank Cass, 1979).

BALIMA, A. S., *Genèse de la Haute-Volta* (Ouagadougou: Presses Africaines, 1969).

BALOGUN, I. A. B., 'A Critical Edition of the *Ihya' al-sunna wa-'ikhmad al-bid'a* of 'Uthman b. Fudi, Popularly Known as Usumanu dan Fodio' (Ph.D. thesis; School of Oriental and African Studies, University of London, 1967).

BATRAN, A. A., 'An Introductory Note on the Impact of Sidi al-Mukhtar al-Kunti (1729–1811) on West African Islam in the 18th and 19th Centuries', *Journal of the Historical Society of Nigeria*, 6/4 (June 1973), 347–52.

—— 'The Kunta, Sidi al-Mukhtar al-Kunti, and the Office of Shaykh al-Ṭarīqa al-Qādiriyya', in J. R. Willis (ed.), *Studies in West African Islamic History*, i. *The Cultivators of Islam* (London: Frank Cass, 1979), 113–46.

—— 'The Qadiriyya-Mukhtariyya Brotherhood in West Africa: The Concept of Tasawwuf in the Writings of Sidi al-Mukhtar al-Kunti (1729–1811)', *Transafrican Journal of History*, 4/1–2 (1974), 41–70.

—— 'Sidi al-Mukhtar al-Kunti and the Recrudescence of Islam in the Western Sahara and the Middle Niger' (Ph.D. thesis; University of Birmingham, 1971).

BECK, L. and KEDDIE, N. R. (eds.), *Women in the Muslim World* (Cambridge, Mass., and London: Harvard University Press, 1980).

BECKER, C., 'Materials for the Understanding of Islam in German East Africa', *Tanzania Notes and Records*, 68 (1968), 31–61.

BICHON, B., 'Les Musulmans de la subdivision de Kombissiri (Haute-Volta)', in id., *Notes et études sur l'Islam en Afrique Noire* (Paris: Recherches et Documents du CHEAM, 1, 1962), 75–102.

BIDMUS, A. K., 'A Literary Appraisal of the Arabic Writings of the Yoruba *Ulama*', M.Phil. thesis (University of Ibadan, 1972).

BLOCH, M., *La Société féodale* (Paris: Albin Michel, 1958); English edn., trans. by Manyon, *Feudal Society* (London: Routledge and Kegan Paul, 1962.)

BOUCHE, D., 'Les Écoles françaises à l'époque de la conquête, 1884–1900', *Cahiers d'études africaines*, 6/2/22 (1966), 228–67.

BOVIN, M., 'Muslim Women in the Periphery: The West African Sahel', in B. Utas (ed.), *Women in Islamic Societies* (London and Malmö: Scandinavian Institute of Asian Studies, 1983), 66–103.

BOYD, A. W., 'To Praise the Prophet: A Processual Symbolic Analysis of "*Maulidi*", a Muslim Ritual in Lamu, Kenya', Ph.D. thesis (Indiana University, 1981).

BOYD, J., 'The Contribution of Nana Asma'u Fodio to the Jihadist Movement of Shehu dan Fodio from 1820 to 1855', M.Phil. thesis (North London Polytechnic, 1982).

—— and LAST, M., 'The Role of Women as *Agents Religieux* in Sokoto', *Canadian Journal of African Studies*, 19/2 (1985), 283–300.

BRENNER, L., 'The Esoteric Sciences in West African Islam', in I. H. Abdalla and B. M. Du Toit (eds.), *African Healing Strategies* (Buffalo, NY: Trado-Medic Books, 1985).

—— *West African Sufi: The Religious Heritage and Spiritual Search of Cerno Bokar Saalif Taal* (London: C. Hurst, 1984).

CAPLAN, A. P., *Choice and Constraint in a Swahili Community* (Oxford: Oxford University Press, 1975).

Cardaire, M., *L'Islam et le terroir africain* (Koulaba: Institut Fondamental d'Afrique Noire (IFAN), Imprimerie du Gouvernement, 1954).

Carré, O. and Dumont, P. (eds.), *Les Radicalismes Islamiques* (Paris: L'Harmattan, 1985).

Christelow, A., 'The Yan Tatsine Disturbances in Kano: A Search for Perspective', *The Muslim World*, 75 (1985).

Clarke, P. B., *West Africa and Islam* (London: Edward Arnold, 1982).

Cohen, A., *Custom and Politics in Urban Africa: A Study of Hausa Migrants in Yoruba Towns* (London: Routledge and Kegan Paul, 1969).

Colvin, L., 'The Shaykh's Men: Religion and Power in Senegambian Islam', in Fisher and Levtzion (eds.), *Rural and Urban Islam in West Africa*.

Constantin, F., *Les Communautés musulmanes d'Afrique orientale* (Pau: CREPAO, 1983).

—— 'Condition féminine et dynamique confrérique en Afrique orientale', *Islam et sociétés au sud du Sahara*, 1 (1987), 58–69.

—— 'Le Saint et le prince: Sur les fondements de la dynamique confrérique en Afrique orientale', *Modes populaires d'action politique*, 3 (1984), 22–47.

Cooper, F., 'Islam and Cultural Hegemony: The Ideology of Slave-Owners on the East African coast', in P. Lovejoy (ed.), *The Ideology of Slavery in Africa* (London: Sage Publications, 1981).

Copans, J., *Les Marabouts de l'arachide* (Paris: Le Sycamore, 1980).

Coulon, C., 'Introduction à une sociologie des confréries en Afrique orientale', *Modes populaires d'action politique*, 3 (1984), 4–21.

—— *Le Marabout et le prince* (Paris: Pedone, 1981).

—— *Les Musulmans et le pouvoir en Afrique noire* (Paris: Karthala, 1983).

Crapanzano, V., *The Hamadsha: A Study in Moroccan Ethno-Psychiatry* (Berkeley: University of California Press, 1973).

Cruise O'Brien, D. B., *The Mourides of Senegal: The Political and Economic Organization of an Islamic Brotherhood* (Oxford: Clarendon Press, 1971).

—— *Saints and Politicians* (Cambridge: Cambridge University Press, 1975).

Cuoq, J., *Les Musulmans en Afrique* (Paris: Maisonneuve et Larose, 1975).

Dekmejian, R. H. and Wyszomirski, M. J., 'Charismatic Leadership in Islam: The Mahadi of the Sudan', *Comparative Studies in Society and History*, 14/2 (Mar. 1972), 193–214.

Deniel, R., *Croyances religieuses et vie quotidienne: Islam et christianisme à Ouagadougou* (Paris and Ouagadougou: CNRS and CVRS, 1970).

Dermenghem, E., *Le Culte des saints dans l'Islam Maghrebin* (Paris: Gallimard, 1954).

DIAWARA, Y., 'Sur la révolte de 3 août 1941 à Bobo-Dioulasso', *Notes et documents voltaïques*, 8/2 (Jan.–Mar. 1975).

DIN-ATTAR, F. U., *Le Mémorial des saints* (Paris: Editions du Seuil, 1976).

DIOP, M. C., 'Les Affaires mourides à Dakar', *Politique africaine*, 1/4 (Nov. 1981).

—— 'La Confrérie mouride: Organisation politique et mode d'implantation urbaine', *Doctorat de troisième cycle* (Université de Lyon II, 1980).

DOI, R., 'The Yoruba Mahdi', *Journal of Religion in Africa*, 4 (1971).

DOKAJI, A., *Kano Ta Dabo Cigari* (2nd edn.; Zaria: Gaskiya, 1978).

DUMONT, F., *L'Anti-Sultan ou al-Hajj Omar Tal du Fouta, combattant de la foi* (Dakar and Abidjan: Les Nouvelles Éditions Africaines, 1974).

DWYER, D. H., 'Women, Sufism and Decision-Making in Moroccan Islam', in L. Beck and N. Keddie (eds.), *Women in the Muslim World* (Cambridge, Mass. and London: Harvard University Press, 1980), 585–98.

ECHENBERG, M. J., 'La Jihad d'Al Kari de Bousse: Un état marka en pays dafing à la fin du XIXe siècle', *Notes et documents voltaïques*, 3 (Apr.–June 1970), 3–43.

EISENSTADT, S. (ed.), *On Charisma and Institution-Building by Max Weber* (Chicago: University of Chicago Press, 1968).

ELBOUDRARI, H., 'Quand les saints font les villes', *Annales ESC* (May–June 1985).

ELGOOD, O., '*Tibb-ul-nabbi* or Medicine of the Prophet', *Osiris*, 14 (1962), 33–192.

EL-MASRI, F. H., 'The Life of Shehu Usuman dan Fodio before the *jihād*', *Journal of the Historical Society of Nigeria*, 11 (1963–4), 435–48.

EL ZEIN, A. H. M., *The Sacred Meadows: A Structural Analysis of Religious Symbolism in an East African Town* (Evanston: Northwestern University Press, 1974).

FERNEA, E. W., *Guests of the Sheikh* (New York: Anchor Books, 1985).

FERNEA, R. A. and E. W., 'Variations in Religious Observance among Islamic Women', in N. R. Keddie (ed.), *Scholars, Saints and Sufis* (Berkeley, London, and Los Angeles: University of California Press, 1972), 385–401.

FERRAT, D., 'La Confrérie des Mourides en France', unpublished *mémoire* (Institut d'Études Politiques d'Aix-en-Provence, 1985).

FIKA, A., *The Kano Civil War and British Over-Rule* (Ibadan: Oxford University Press, 1978).

FISHER, H. J., 'Prayer and Military Activity in the History of Muslim Africa South of the Sahara', *Journal of African History*, 12/3 (1971).

—— and LEVTZION, N. (eds.), *Rural and Urban Islam in West Africa*, published as *African and Asian Studies*, 20/1 (Mar. 1986).

GBADAMOSI, T. G. O., *The Growth of Islam among the Yoruba, 1841–1908* (London: Longman, 1978).

GEERTZ, C., 'Centres, Kings and Charisma: Reflections on the Symbolics of Power', in *Local Knowledge* (New York: Basic Books, 1983).

GELLNER, E., *Muslim Society* (Cambridge: Cambridge University Press, 1981).

—— 'Pouvoir politique et fonction religieuse dans l'Islam marocain', *Annales: Économies, sociétés, civilisations*, 3 (May–June 1970), 699–713.

—— *Saints of the Atlas* (London: Weidenfeld and Nicholson, 1969).

GILSENAN, M., *Recognizing Islam* (London: Croom Helm, 1982).

GNANKAMBARY, B., 'La Révolte bobo de 1916 dans le cercle de Dedougou', *Notes et documents voltaïques*, 11/3–4 (Apr.–Sept. 1978), 1–38.

GUMI, A. M., *Tarjamar Ma' anonin Alkur' ani Mai Girma* (Beirut: Dar al-Arabia, 1982).

GURSOY-NASKALI, E. E., 'Women Mystics in Islam', in B. Utas (ed.), *Women in Islamic Societies*, pp. 238–44.

HAMET, I., 'Littérature arabe saharienne', *Revue du Monde musulman*, 12 (1910), 196–213, 380–405.

HAMMOUDI, A., 'Sainteté, pouvoir et société: Tamergrout aux XVII^e^ et XVIII^e^ siècles', *Annales ESC*, Paris (1980), 615–41.

HARRISON, C., *France and Islam in West Africa* (Cambridge: Cambridge University Press), forthcoming.

HISKETT, M., *The Development of Islam in West Africa* (London: Longman, 1984).

—— *The Sword of Truth: The Life and Times of the Shehu Usuman dan Fodio* (New York: Oxford University Press, 1973).

HJORT, A., *Savanna Town: Rural Ties and Urban Opportunities in Northern Kenya* (Stockholm: Stockholm Studies in Social Anthropology, 1979).

HOUDAS, O. (ed. and trans.), *Ta'rīkh es-Soudan* (Paris: A-Maisonneuve, 1900).

—— (ed. and trans.), *Ta'rīkh al-Fattash* (Paris: A-Maisonneuve, 1913).

HUNWICK, J. O., 'Neo-Hanbalism in Southern Nigeria: The Reformist Ideas of al-Hajj Adam al-Illuri of Agege', in id., *Islam et sociétés au sud du Sahara—notes et documents* (Paris: Éditions de la Maison des Sciences de l'Homme, 1, May 1987), 18–27.

—— *Sharī'a in Songhay: The Replies of al-Maghili to the Questions of Askia al-Hajj Muhammad* (London: Oxford University Press with the British Academy, 1985).

IZARD, M., *Introduction à l'histoire des royaumes Mossi* (Paris and Ouagadougou: CNRS and CVRS, 1970).

JACKSON, R. H. and ROSBERG, C. (eds.), *Personal Rule in Black Africa* (Berkeley: University of California Press, 1982).

KABA, L., *The Wahhabiyya: Islamic Reform and Politics in West Africa* (Evanston: Northwestern University Press, 1974).

KEDDIE, N. R. (ed.), *Scholars, Saints and Sufis* (Berkeley, London, and Los Angeles: University of California Press, 1972).

KEPPEL, G., *Le Prophète et Pharaon: Les mouvements islamistes dans l'Égypte contemporaine* (Paris: La Découverte, 1983); English edition, trans. by J. Rothschild, *The Prophet and Pharaoh: Muslim Extremism in Egypt* (London: Al-Saqi Books, 1985).

KLEIN, M., *Islam and Imperialism in Senegal* (Edinburgh: Edinburgh University Press, 1968).

KOUANDA, A., 'Les Yarse: Fonction commerciale, religieuse et légitimité culturelle dans le pays moaga', *Doctorat de troisième cycle* (UER d'Histoire, Université de Paris I, 1984).

LAST, M., 'From Sultanate to Caliphate: Kano 1450–1800', in B. M. Barkindo (ed.), *Studies in the History of Kano* (Ibadan: Heinemann, 1983), 67–91.

—— 'Historical Metaphors in the Kano Chronicle', *History in Africa*, 7 (1980), 161–78.

—— 'A Note on Attitudes to the Supernatural in the Sokoto *jihād*', *Journal of the Historical Society of Nigeria*, 4/1 (Dec. 1967), 3–13.

—— *The Sokoto Caliphate* (London: Longman, 1967).

LAVERS, J. E., 'Diversions on a Journey', in Y. F. Hasan and P. Doornbos (eds.), *The Central Bilad al-Sudan* (Khartoum: El-Tammadon Press, 1977), 216–32.

LE GUENNEC COPPENS, F., *Femmes voilées de Lamu (Kenya)* (Paris: Éditions Recherche sur les Civilisations, 1983).

LEVTZION, N., *Ancient Ghana and Mali* (London: Methuen, 1973).

—— and VOLL, J. O. (eds.), *Eighteenth-Century Renewal and Reform in Islam* (Syracuse: Syracuse University Press, 1987.

LUBECK, P., *Islam and Urban Labour in Kano* (Cambridge: Cambridge University Press, 1987).

MAGASSOUBA, M., *L'Islam au Sénégal: Demain les mollahs?* (Paris: Éditions Karthala, 1985).

MAHIBOU, S. N. and TRIAUD, J. L. (eds. and trans.), *Voilà ce qui est arrivé:* Bayan ma Waqaʿa *d'al Hajj ʿUmar al-Futi, Plaidoyer pour une guerre sainte en Afrique de l'Ouest au 19e siècle* (Paris: Éditions du CNRS, 1983).

MARTIN, B. G., *Muslim Brotherhoods in Nineteenth-Century Africa* (Cambridge: Cambridge University Press, 1976).

—— 'Muslim Politics and Resistance to Colonial Rule: Shaykh Uways b. Muhammad al-Barawi and the Qādiriyya Brotherhood in East Africa', *Journal of African History*, 10/3 (1969), 471–86.

—— 'Unbelief in the Western Sudan: 'Uthman dan Fodio's "Ta'līm al-ikhwān"', *Middle Eastern Studies*, 4 (1967), 50–97.

MEILLASSOUX, C. (ed.), *The Development of Indigenous Trade and Markets in West Africa* (Oxford: Oxford University Press and International African Institute, 1971).

MERNISSI, F., *Sexe, idéologie, Islam* (Paris: Éditions Tierce, 1983).

MOREAU, R. L., 'Les Marabouts de Dori', *Archives de sociologie religieuse*, 17 (Jan.–June 1964), 113–34.

MUHAMMAD BELLO, *Infāq al-maisūr fī ta'rīkh bilād al-takrūr*, ed. by C. E. J. Whitting (London: Luzac, 1957).

MUSALLAM, B. S., *Sex and Society in Islam* (Cambridge: Cambridge University Press, 1983).

NICOLAS, G., *Dynamique de l'Islam au sud du Sahara* (Paris: Presses Orientalistes de France, 1981).

NIMTZ, A. H., *Islam and Politics in East Africa: The Sufi Order in Tanzania* (Minneapolis: University of Minnesota Press, 1980).

OFORI, P. E., *Islam in Africa South of the Sahara: A Select Bibliographic Guide* (Nendeln: KTO Press, 1977).

OTAYEK, R., 'La Crise de la communauté musulmane de Haute-Volta: L'Islam voltaïque entre réformisme et tradition, autonomie et subordination', *Cahiers d'études africaines*, 24/3/95 (1984), 299–320.

PADEN, J. N., *Ahmadu Bello: Sardauna of Sokoto* (London: Hodder and Stoughton, 1986).

—— *Religion and Political Culture in Kano* (Berkeley: University of California Press, 1973).

PALMER, H. R., *Sudanese Memoirs* (Lagos: Government Printer, 1928).

POPOVIC, A. and VEINSTEIN, G. (eds.), *Les Ordres mystiques dans l'Islam: Cheminements et situation actuelle* (Paris: EHESS, 1986).

POUWELS, R., *Horn and Crescent: Cultural Change and Traditional Islam on the East African Coast, 800–1900* (Cambridge: Cambridge University Press, 1987).

—— 'The Medieval Foundations of East African Islam', *International Journal of African Historical Studies*, 11/2 (1978), 201–26; and 11/3 (1978), 393–409.

QUIMBY, L., 'Islam, Sex Roles and Modernization in Bobo-Dioulasso', in B. J. Rossette (ed.), *The New Religions of Africa* (Norwood, NJ: Ablex Publishing Corp., 1977), 203–18.

RIESMAN, P., *Société et liberté chez les Peuls Djelgobé de Haute-Volta: Essai d'anthropologie introspective* (Paris and La Haye: Mouton, 1974); English edition: *Freedom in Fulani Social Life* (Chicago and London: University of Chicago Press, 1977).

ROBINSON, D., *The Holy War of Umar Tal: The Western Sudan in the Mid-Nineteenth century* (Oxford: Clarendon Press, 1985).

SCHILDKROUT, E., 'Dependence and Autonomy: The Economic

Activities of Secluded Hausa Women in Kano, Nigeria', in E. G. Bay (ed.), *Women and Work in Africa* (Boulder, Colo.: Westview Press, 1982), 429–50.

Séguy, J., 'Charisme, sacerdoce, fondation: Autour de L. M. Grignion de Montfort', *Social Compass*, 29/1 (1982), 5–24.

Skinner, E. P., 'The Diffusion of Islam in an African Society', *Annals of the New York Academy of Science*, 96/2 (Jan. 1962), 659–69.

—— 'Islam in Mossi Society', in I. M. Lewis (ed.), *Islam in Tropical Africa* (London: International African Institute and Hutchinson University Library for Africa, 1980), 173–93.

—— *Les Mossi de la Haute-Volta* (Paris: Les Éditions Internationales, 1964).

Stewart, C. C., with Stewart, E. E., *Islam and Social Order in Mauritania* (Oxford: Clarendon Press, 1973).

Strobel, M., *Muslim Women in Mombasa, 1880–1975* (New Haven, Conn., and London: Yale University Press, 1979).

Traoré, A., *Islam et colonisation en Afrique: Cheikh Hamahoullah, homme de foi et résistant* (Paris: Maisonneuve et Larose, 1983).

Traoré, B., 'Le Processus d'islamisation à Bobo-Dioulasso jusqu'à la fin du xixe siècle: Approche historique et sociologique', Mémoire de maîtrise (Université de Ouagadougou, 1984).

Triaud, J. L. and Constantin, F. (eds.), *Les Leaders religieux islamiques en Afrique Orientale* (Paris: Credu/Karthala, forthcoming).

Trimingham, J. S., *A History of Islam in West Africa* (London: Oxford University Press, 1962).

—— *The Influence of Islam upon Africa* (London and New York: Longman and Librarie du Liban, 1980).

—— *Islam in East Africa* (Oxford: Clarendon Press, 1962).

—— *Islam in Ethiopia* (London: Oxford University Press, 1952).

—— *Islam in the Sudan* (Oxford: Clarendon Press, 1949).

—— *Islam in West Africa* (Oxford: Clarendon Press, 1959).

—— *The Sufi Orders in Islam* (Oxford: Clarendon Press, 1971).

Turner, B. S., *Weber and Islam* (London: Routledge and Kegan Paul, 1974).

Tyam, M. A., *La Vie d'El Hadj Omar, Qacida en Poular*, ed. and trans. by H. Gaden (Paris: Institut d'Ethnologie, 1935).

Usman, Y. B. (ed.), *Studies in the History of the Sokoto Caliphate: The Sokoto Seminar Papers* (Zaria: Department of History, Ahmadu Bello University, for the Sokoto State History Bureau, 1979).

Utas, B. (ed.), *Women in Islamic Societies* (London and Malmö: Scandinavian Institute of Asian Studies, 1983).

ʿUthman b. Fudi, *Bayān wujūb al-hijra ʿalā 'l-ʿibād*, ed. and trans. from the Arabic by F. H. El-Masri (Oxford: Oxford University Press, 1978).

—— *Wird* (Zaria: Gaskiya Corporation, 1962).

WEBER, M., *Economy and Society*, ed. G. Roth and C. Wittich (New York: Bedminster, 1968).

WILLIS, J. R. (ed.), *Slaves and Slavery in Muslim Africa*, 2 vols. (London: Frank Cass, 1985).

—— *Studies in West African Islamic History*, i. *The Cultivators of Islam* (London: Frank Cass, 1979).

ZAHRADEEN, M. S. 'The Place of Mosques in the History of Kano', in B. M. Barkindo (ed.), *Studies in the History of Kano* (Ibadan: Heinemann, 1983).

ZOGHBY, S. M., *Islam in Sub-Saharan Africa: A Partially Annotated Guide* (Washington, DC: Library of Congress, 1978).

# Glossary

*ʿālim*: scholar, learned man, or teacher (plural: *ʿulamāʾ*).
*asnād*: chain or link, as in academic genealogy.
*baraka*: numinous authority, blessing grace—transmitted by holy men and holy places.
*dhikr*: the recitation of prayers which recall the name of Allah; a spiritual exercise.
*fuqahāʾ*: jurists, conventionally contrasted with Sufis.
*ijāza*: certificate proving competence in a text and permission to pass on that teaching to others.
*ʿilm*: knowledge; science.
*ijtihād*: reinterpretation of Law.
*jihād*: action, war for the Faith.
*karāmāt*: miracles.
*khalwa*: solitary withdrawal for prolonged meditation.
*madrasa*: school, often associated with a mosque.
*mallam*: scholar (the Hausa form of *muʿallim*); cf. *ʿālim*.
*murīd*: novice, aspirant; disciple (plural: *murīdūn*).
*quṭb*; *quṭb al-aqṭāb*: 'Pole'; 'Pole of poles'—the highest level in the Sufi hierarchy of saints.
*shaikh*: a title indicating, for Sufis, high spiritual attainment, and the highest level of leadership in a Sufi order.
*sharīʿa*: Islamic law.
*silsila*: academic or spiritual 'genealogy'; chain of authority for the transmission of a text or a *wird* (plural: *salāsil*).
*ṣūfī*: an Islamic mystic.
*ṭālib*: student or disciple (plural: *ṭalaba*).
*ṭarīqa*: a way, path; a Sufi brotherhood or order (plural: *ṭuruq*).
*ʿulamāʾ*: plural of *ʿālim*.
*walī*: 'friend of Allah'; saint (plural: *awliyāʾ*).
*walāya*: in Sufi terminology, sainthood; the sovereignty or protection given by Allah to the faithful. In conventional speech it means power, government.
*wird*: the litany of special prayers peculiar to a Sufi order.
*zāwiya*: Sufi centre or meeting-place; residence of scholars and students (plural: *zawāyā*).
*ziyāra*: visit (to a shrine).

# Index